Ethological and physiological indicators of positive emotions in juvenile pigs
(*Sus scrofa*)

I0131739

Ethological and physiological indicators of positive emotions in
juvenile pigs (*Sus scrofa*)

Dissertation

for the Doctoral Degree

at the Faculty of Agricultural Sciences,

Georg-August University Göttingen

presented by

Lisa Christine McKenna

born in Pirmasens

Göttingen, February 2018

Bibliografische Information der Deutschen Nationalbibliothek

Die Deutsche Nationalbibliothek verzeichnet diese Publikation in der
Deutschen Nationalbibliografie; detaillierte bibliographische Daten sind im Internet
über http://dnb.d-nb.de abrufbar.
1. Aufl. - Göttingen: Cuvillier, 2018
   Zugl.: Göttingen, Univ., Diss., 2018

D7

1st Referee:	Prof. Dr. Martina Gerken
Co-Referee:	Prof. Dr. Ute Knierim
Date of disputation:	15th February 2018

© CUVILLIER VERLAG, Göttingen 2018
   Nonnenstieg 8, 37075 Göttingen
   Telefon: 0551-54724-0
   Telefax: 0551-54724-21
   www.cuvillier.de

   ISBN 978-3-7369-9864-3
   eISBN 978-3-7369-8864-4

TABLE OF CONTENTS

SUMMARY

The natural behavior of pigs is adapted to very complex environments and is characterized by a high proportion of exploration behaviour. In modern intensive husbandry systems, the behaviour pigs would show in natural environments cannot be expressed to the degree that is necessary for the animals to maintain a positive affective state. Health and behaviour problems arise in intensive pig husbandry, leading to stress and negative emotional states in the animals. This animal welfare problem fuels growing concerns of the public. As negative emotional states, such as stress, are more intensely expressed and potentially life threatening, they are easier to study. The expression of positive emotion however is more subtle, but no less important. If situations in which positive emotions arise can be identified, and incorporated into livestock production systems, animal welfare can be increased to a large extent. Deliberate actions the animal can take upon facing a challenge within its environment which enable a positive outcome facilitate controllability and therefore positive emotion (Boissy et al., 2007). In literature, different factors contributing to the well-being of animals in livestock husbandry have been studied. In this context, for example anticipation behaviour (Spruijt et al., 2001; Dudink et al., 2006), environmental and cognitive enrichment (Puppe et al., 2007; Martin et al., 2015) as well as the measurement of physiological indicators help to measure positive emotional states in livestock animals (Chapter I).

This study was therefore conducted with the objective to identify, elicit and measure behavioural and physiological indicators of positive emotion in pigs. Therefore three experiments were carried out on growing pigs.

In the first experiment (Chapter II), it was hypothesized that objects and materials of more complex nature would elicit more exploratory behaviour and an increase in parasympathetic influence in the cardiovascular system of pigs. Therefore, 18 growing pigs (9 female, 9 male) were housed in substrate enriched environments and tested in novel object tests. The animals were led to the test room in fixed groups of three animals to avoid social isolation stress. In the test room, the animals were fitted with devices measuring heart rate (HR) and heart rate variability (RR, RMSSD) and led to the test arena where three of the same objects were laid out for them. The animals were exposed to the novel objects for 7 mins. The objects which were used in the novel object test were: a heap of soil, rubber dog toys filled with grapes (Kong®), IQ games filled with grapes, examination gloves filled with water, ropes, rubber ducks, dried leaves and a wallow (a wooden frame lined with pond-liner and filled with soil and water). In this experiment, the effects of gender and object on the exploratory, play and tail wagging behaviour and the corresponding cardiovascular parameters of the animals were measured. Gender did not affect any of the

7

behavioural parameters as well as HR and RR. Female animals however, had larger RMSSD values than male animals. In addition, an interaction effect between gender and object was found. Male animals explored the food-baited IQ and Kong® object more, while females preferred to explore the wallow. Object significantly affected all behaviour parameters. Most explored objects were the soil, the wallow and the dried leaves (77.6 ± 2.8, 67.8 ± 2.5 and 89.6 ± 2.3% of the time, respectively). Least explored objects were the rubber ducks, the ropes and the gloves (31.4 ± 2.6, 24.0 ± 2.7 and 19.3 ± 2.1%, respectively). Play was shown most during the tests with the novel objects rope, Kong® and duck. Tail wagging behaviour was shown mostly with duck and wallow. HR was highest during the tests with the ducks (169.8 ± 2.9 bpm), and lowest during the tests with the leaves (149.8 ± 2.3 bpm). This result was mirrored by highest RR with the leaves (405.9 ± 5.9 ms) and lowest RR with the duck (350.7 ± 7.4 ms). Across all novel objects, non-significant Kendalls Tau correlation coefficients (τ) between explorative behaviour and HR, RR and RMSSD were found (τ = -0.020, -0.024 and -0.007, respectively). These findings show that complex objects are more suitable as occupational materials for pigs than rigid objects. The relations between exploratory behaviour and cardiovascular parameters were not consistent across all tested objects. Cardiovascular parameters related to emotional states are also confounded with locomotor activity and diurnal rhythm. Together with the limited number of analysable cardiovascular data, this might have caused the partial incoherence of the results in this study.

Pigs housed in stimulus and substrate enriched environments could display species specific behaviour more frequently and it is assumed that they experience positive affective states more frequently. We hypothesized that the previous enrichment experience would also affect their behaviour in a test situation. The objective of the second experiment (Chapter III) was to find out whether variation of enrichment in housing environment would alter the reactions toward novel objects in a test situation. 36 female growing pigs were housed in three different housing environments (n = 12 animals in each environment). The three housing environments were: 1. non-enriched (NE), 2. enriched with substrate and straw (E) and 3. enriched with substrate, straw and weekly changing extra stimuli i.e. grass cut, twigs, vegetables, hay (SE). Animals of each environment were divided into a) experimental animals, which were confronted with novel objects in the test arena and b) controls, which were confronted with an open-field situation (empty test arena). The novel objects used for the experimental animals were soil, wallow, Kongs® and rubber ducks. In this experiment, the effects of treatment, environment and object on the play and tail wagging behaviour and the corresponding cardiovascular parameters of the animals were measured. Play and tail wagging behaviour were additionally summarized as positive emotion score (PES). Results show that the cardiovascular parameters were not significantly influenced by environment or treatment. The experimental animals showed more tail wagging and play behaviour as well as a higher PES when the wallow was tested (1.0 ± 4.8, 6.0 ± 4.8 and 3.5 ± 1.0, respectively) compared to the rubber ducks (0.0± 1.4, 0.0 ± 0.3 and 0.3 ± 0.8, respectively). Housing environment had an influence on tail wagging behaviour. Here, animals of the NE environment showed more tail wagging behaviour than animals of the E environment (NE: 2.1 ± 3.9, E: 0.8 ± 2.4). Similarly, PES was significantly affected by housing environment (NE = 227, E = 113 and SE = 121, respectively). The main influencing effect in this experiment was the treatment. Treatment had a significant effect on all behavioural traits recorded with higher values in experimental animals for tail wagging (2.1 vs 0.4), play (3.4 vs 0.5) and PES (5.5 vs 0.9) than the controls. HR tended to be higher in experimental animals than in controls (180.1 vs 171.4, p=0.07), while RR showed a reversed tendency (337.4 vs 353.9, P=0.06). Treatment had no significant effect on RMSSD. The environment x treatment interactions were not significant either. We conclude from these findings that novel objects can induce a positive emotional state expressed by increased play and tail wagging behaviour. High locomotor activity during the exploration of the novel objects and test arena might have superimposed the effects of treatment and housing environment on the cardiovascular parameters. The higher tail wagging behaviour in NE animals could be interpreted as a rebound effect, reflecting the gap between their unsatisfied need for exploration in the barren home environment and the possibility to do so in the test setting. Regular confrontation with novel objects might therefore induce positive emotional states in pigs.

The focus of the third experiment (Chapter IV) was the influence of different levels of environmental enrichment in housing environment on cognitive abilities in young pigs. We hypothesized, that pigs reared in an environment containing more stimuli would show more cognitive flexibility which would enable them to solve a cognitive task faster than conspecifics housed in an impoverished environment. The animals and housing environments were the same as in the second experiment. The animals were tested in two cognitive tasks: 1. Board: A wooden board was angled at 45°. A food item (grape) was then attached to the end of 1m long piece of twine, and placed underneath the angled board. The task was solved once the animal managed to retrieve the food item from under the board by pulling the rope or tilting and pushing away the board. 2. Pipe: A plastic pipe was attached to a wooden board at 90°. Again, a food item (grape) was fixed to the end of a 1m long piece of twine and placed into the pipe from above. The task was considered solved as soon as the test animal retrieved the food item from out of the pipe by pulling the rope or tilting and pushing away the board. All animals were tested twice per cognitive task. The effects of treatment, environment and trial were tested on the latency to first touch the test apparatus, the latency to solve the task, the latency from first touching to solving the task and the number of tasks solved. Results show that animals from the NE environment tended to touch the test apparatus faster in the board task, which was tested first. This finding was not evident for the pipe task, the second test. Animals from the NE environment also tended to solve the pipe task faster than animals from enriched environments. However, the proportion of solved tasks did not differ between environments for both tasks. Trial influenced the latency until solving the task with animals solving the task quicker in the second trial. The board task was solved significantly faster by the experimental animals (who were tested in novel object tests in the previous experiment) compared to the controls (who were not used to encountering objects of any kind in the test arena). In the second trial of the pipe task, experimental animals showed reduced latency to solve the task compared to controls. This could be due to their previous experience with the novel objects. We conclude from our findings that the level of enrichment in the home environment had a tendential effect on the motivation of pigs to explore and interact with the test apparatus used in this study. The pigs coming from the non-enriched home environment tended to solve the cognitive tasks faster, which is probably not due to differences in cognitive capacities but rather a rebound effect in motivation to show species specific exploration behaviour. Also, the previous experience with the novel objects might have made the experimental animals bolder in their approach to the cognitive tasks. The novelty of the situation could have led the control sows to explore the test apparatus more cautiously and therefore also solve the task slower than the test animals.

The ethological and physiological component of emotion is measurable. The third component of emotion, the conscious experience of the emotional state however, is not directly measurable in animals. While humans have means of verbal transfer to account for felt experiences, this is not possible with animals. The measured behavioural and physiological changes in animals are interpreted as positive, but it remains unclear what the animals actually feel (Chapter V). Measurable outcomes of this study showed that cardiovascular parameters are indeed promising indicators of potentially positive emotional states. However, they are easily confounded and larger data quantities are necessary for further exploration of their association to ethological expressions of positive emotion. Play and tail wagging behaviour but also exploratory behaviour is a potential behavioural expression of positive emotional states indicated by the pigs in this study. Lastly, living environment can alter reactions of animals in behavioural tests. Especially housing in impoverished conditions is compromising the pigs need for exploration, altering the animals' reactions in behaviour tests (Chapter V).

ZUSAMMENFASSUNG

Das arteigene Verhalten von Schweinen ist an eine sehr komplexe Umwelt ange-passt und zeichnet sich durch ein hohes Maß an explorativem Verhalten aus. In den modernen Systemen der landwirtschaftlichen Nutztierhaltung, ist es Schweinen nur sehr eingeschränkt möglich ihr natürliches Verhalten in ausreichendem Maß auszule-ben. Dies verursacht Stress, und es kommt zu gesundheitlichen und verhaltensbiolo-gischen Problemen, denen negative emotionale Zustände zugrunde liegen können. Diese Tierwohl-Problematik ist ebenfalls von wachsendem öffentlichem Interesse. Negative emotionale Zustände werden durch Verhalten stärker ausgedrückt und sind daher einfacher zu messen als positive emotionale Zustände, die subtiler ausge-drückt werden, jedoch in Bezug auf Tierwohl nicht weniger wichtig sind. Sobald Situ-ationen, welche positive emotionale Zustände gezielt auslösen können identifiziert, und in tägliche Abläufe in der Tierhaltung eingebracht werden können, könnte dies zur Steigerung des Tierwohls beitragen. Mit Herausforderungen der Umwelt umge-hen zu können, einen aktiven Einfluss darauf zu haben und deren Ausgang positiv beeinflussen zu können, fördert die Kontrollierbarkeit der Umwelt des Tieres und so-mit auch damit einhergehende positive emotionale Zustände (Boissy et al., 2007). Bisherige Studien konnten unterschiedliche Faktoren benennen, welche das Wohlbe-finden von Tieren in der landwirtschaftlichen Nutztierhaltung messen können. In die-sem Zusammenhang kann beispielsweise die Untersuchung von Erwartungshaltun-gen (Spruijt et al., 2001; Dudink et al., 2006), Umweltanreicherung und kognitiver An-reicherung (Puppe et al., 2007; Martin et al., 2015) oder die Messung physiologischer Indikatoren helfen, emotional Zustände landwirtschaftlicher Nutztiere zu messen (Ka-pitel I). Die vorliegende Studie wurde angefertigt um verhaltensbiologische und phy-siologische Indikatoren positiver emotionaler Zustände beim Schwein zu identifizie-ren, hervorzurufen und zu messen. Dafür wurden drei Experimente durchgeführt.

Die erste Untersuchung (Kapitel II) erfolgte an 18 jungen Schweinen (9 weibliche und 9 männliche Tiere). Alle Tiere wurden nach Geschlecht getrennt in angereicherter Umwelt gehalten und in Novel-Object-Tests getestet. Die Versuchshypothese hier war, dass Objekte und Materialien komplexerer Natur mehr exploratives Verhalten bei den Tieren auslösen und zu einem Anstieg parasympathischer Anteile des kar-diovaskulären Systems führen. Zur Durchführung der Verhaltenstests wurden die Tiere in festen Dreiergruppen in einen Testraum geführt. Dort wurden die Herzfre-quenzmessgeräte angebracht, welche die Herzrate (HR) und die Herzfrequenzvaria-bilität (RR, RMSSD) aufzeichneten. In der Testarena wurden drei unbekannte Ob-jekte ausgelegt und die Tiere hatten die Möglichkeit die Objekte für die Dauer von 7 Minuten zu erkunden. Die Objekte die für die Studien genutzt wurden waren: aufge-häufte Erde, mit Trauben gefüllte Hundespielzeuge aus Gummi (Kong®), mit Trauben gefüllte Intelligenzspiele, mit Wasser gefüllte Gummihandschuhe, Hanfseile, Gummienten, aufgehäuftes Laub und eine Suhle (ein Holzrahmen wurde dazu mit Teichfolie ausgekleidet und mit Wasser und Erde gefüllt). In diesem Experiment wurde der Effekt des Geschlechts der Tiere und des getesteten Objektes auf das Erkundungs-, Spiel- und Schwanzwedelverhalten und die korrespondierenden kardiovaskulären Parameter gemessen. Der Effekt des Geschlechts der Tiere hatte keinen signifikanten Einfluss auf die verhaltensbiologischen oder kardiovaskulären Parameter HR und RR. RMSSD Werte der weiblichen Tiere allerdings waren höher als die der männlichen Tiere. Ein Interaktionseffekt zeigte weiterhin, dass die männlichen Tiere die mit Futter gefüllten Objekte mehr erkundeten, während die weiblichen Tiere die Suhle bevorzugten. Das getestete Objekt zeigte einen signifikanten Einfluss auf alle Verhaltensparameter. Objekte, die am meisten erkundet wurden, waren die Erde, die Suhle und das Laub (77.6 ± 2.8, 67.8 ± 2.5 und 89.6 ± 2.3% der Zeit, dementsprechend). Am wenigsten erkundet wurden die Gummienten, die Hanfseile und die Gummihandschuhe (31.4 ± 2.6, 24.0 ± 2.7 und 19.3 ± 2.1%, dementsprechend). Spielverhalten wurde am meisten gezeigt wenn Hanfseile, Kong® und Gummienten getestet wurden während Schwanzwedeln am meisten bei der Suhle und den Gummienten beobachtet werden konnte. Die höchste HR und niedrigste RR wurde während der Tests mit den Gummienten gemessen (HR: 169.8 ± 2.9 Schläge pro Minute, RR: 350.7 ± 7.4 Millisekunden), und die geringste HR und höchste RR zeigten die Tiere beim Test mit Laub (HR: 149.8 ± 2.3 Schläge pro Minute, RR: 405.9 ± 5.9 Millisekunden). Über alle getesteten Objekte hinweg wurden keine signifikanten Kendalls Tau Korrelationskoeffizienten (τ) zwischen Erkundungsverhalten und HR, RR oder RMSSD gefunden (τ = -0.020, -0.024 und -0.007, dementsprechend). Diese Ergebnisse zeigen, dass sich komplexere Objekte und Materialien besser zur Beschäftigung von Schweinen eignen als Objekte mit einfacherer Struktur. Die Beziehung zwischen Erkundungsverhalten und kardiovaskulären Parametern war in Ansätzen sichtbar, jedoch nicht kohärent nachweisbar für alle Objekte. Kardiovaskuläre Parameter sind als Hinweis auf emotionale Zustände oft von Einflüssen wie beispielsweise körperlicher Bewegung, Aktivität oder Tagesrhythmus überlagert. In Kombination mit der geringen Menge an kardiovaskulären Daten welche für diese Untersuchung zur Verfügung standen, könnten diese Faktoren die teilweise Inkohärenz der Ergebnisse verursacht haben.

Schweine die in substrat-angereicherten Haltungssystemen gehalten werden, können arteigenes Verhaltens ausgeprägter ausleben. Es wäre denkbar, dass diese Tiere daher öfter positive emotionale Zustände erleben. Die Hypothese der zweiten Untersuchung war, dass das Level der Anreicherung eines Haltungssystems potentiell das Verhalten von Tieren in einer Testsituation beeinflusst (Kapitel III). Der Effekt einer Variation der Anreicherung in der Haltungsumwelt (von karg bis stark angereichert) auf die Verhaltensreaktionen von 36 weiblichen Schweinen wurde hier in einem Novel-Object Test untersucht. Die drei Haltungsumwelten der jeweils 12 Tiere waren: 1. Nicht angereichert (NE), 2. mit Stroh und Spänen angereichert (E) und 3. Mit Stroh, Spänen und zusätzlich wechselnden Stimuli (Grasschnitt, Äste, Gemüse, Heu) angereichert (SE). Die Tiere wurden innerhalb jeder Haltungsumwelt jeweils in Testtiere und Kontrolltiere unterteilt. Während die Testtiere im Testbereich mit den unbekannten Objekten konfrontiert wurden, fanden die Kontrolltiere den Testbereich leer vor. Die verwendeten unbekannten Objekte waren Blumenerde, eine Suhle, Kongs® und Gummi-Enten. In dieser Studie wurden die Effekte der Behandlung, der Haltungsumwelt und des Objektes auf das Spiel- und Schwanzwedel-Verhalten sowie der Herzfrequenz und der Herzfrequenzvariabilität der Tiere gemessen. Spiel und Schwanzwedel-Verhalten wurden zusätzlich als „Positive Emotion Score" (PES) zusammengefasst. Es konnte gezeigt werden, dass die kardiovaskulären Parameter weder von der Haltungsumwelt, noch von der Behandlung signifikant beeinflusst wurden. Die Testtiere zeigten mehr Schwanzwedeln und Spielverhalten, sowie eine höhere PES bei der Suhle (1.0 ± 4.8, 6.0 ± 4.8 und 3.5 ± 1.0, entsprechend) verglichen mit den Gummi-Enten (0.0 ± 1.4, 0.0 ± 0.3 und 0.3 ± 0.8, entsprechend). Die Tiere aus der NE Umwelt zeigten signifikant mehr Schwanzwedeln als die Tiere aus der E Umwelt (NE: 2.1 ± 3.9, E: 0.8 ± 2.4). Die PES wurde in ähnlicher Weise von der Haltungsumwelt geprägt (NE = 227, E = 113 und SE = 121). Die Behandlung der Tiere beeinflusste alle Verhaltensparameter durch höhere Werte der Testtiere verglichen mit den Kontrolltieren (Schwanzwedeln: 2.1 vs 0.4, Spiel: 3.4 vs 0.5 und PES: 5.5 vs 0.9). Der Effekt der Behandlung auf die kardiovaskulären Daten zeigte tendenziell höhere HR bei den Testtieren verglichen mit den Kontrolltieren (180.1 vs 171.4, p=0.07), und eine gegenläufige RR Tendenz (337.4 vs 353.9, P=0.06). RMSSD wurde von der Behandlung nicht beeinflusst. Die Interaktionen zwischen Haltungsumwelt und Behandlung zeigten ebenfalls keine Effekte. Aus diesen Ergebnissen kann geschlussfolgert werden, dass unbekannte Objekte bei Schweinen Verhaltensweisen auslösen, die auf positive emotionale Zustände hinweisen (vermehrtes Spiel und Schwanzwedel-Verhalten). Eine erhöhte Bewegungsaktivität während der Erkundung der Objekte könnte die Effekte der Behandlung und der Haltungsumwelt bezüglich der Herzparameter überdeckt haben. Die höheren PES Werte und das vermehrte Schwanzwedeln der Tiere aus der NE Umwelt weist möglicherweise auf einen sogenannten „Rebound-Effekt" hin. Dieser reflektiert das unbefriedigte Bedürfnis arteigenes Erkundungsverhalten in der Haltungsumwelt zu zeigen, welches in der Testsituation dann durch erhöhte Reaktivität kompensiert wird. Die regelmäßige Konfrontation von Schweinen mit unbekannten Objekten könnte daher zu einer Steigerung der positiven Emotionen und somit des Wohlbefindens der Tiere beitragen.

Der Fokus des dritten Experiments (Kapitel IV) war der Einfluss unterschiedlicher An-reicherungslevel in der Haltungsumwelt auf kognitive Fähigkeiten junger Schweine. Die Hypothese dieser Untersuchung war, dass die kognitive Flexibilität von Schwei-nen aus stark angereicherter Umwelt höher ist und diese somit eine kognitive Auf-gabe schneller als ihre Artgenossen aus karger Umwelt lösen können. Dazu wurden dieselben Tiere und Haltungsumwelten wie schon in der vorherigen Untersuchung (Kapitel III) herangezogen. Es wurden zwei kognitive Aufgaben getestet: 1. Brett: Ein Futterstück (Traube) wurde an einer Schnur (1m) unter einem im 45° Winkel ge-schrägten Brett versteckt. Die Aufgabe galt als gelöst sobald das Tier das Futter durch ziehen an der Schnur oder durch Wegdrücken des Brettes erreichte. 2. Rohr: Ein Plastikrohr wurde im 90° Winkel an einem Holzbrett befestigt. Auch hier wurde ein Futterstück (Traube) an einer Schnur (1m) in das Rohr gehängt. Die Aufgabe war gelöst sobald das Tier das Futter durch ziehen an der Schnur oder durch Kippen des Brettes erreichte. Jedes Tier wurde zwei Mal an beiden Aufgaben getestet (zwei Testdurchgänge). Der Einfluss der Haltungsumwelt, der Behandlung in der vorheri-gen Untersuchung und des Durchgangs wurde auf die folgenden Parameter getestet: Latenzzeit bis zur Erstberührung des Testapparates, Latenzzeit bis zur Lösung der Aufgabe, deren Differenz (Zeit von Erstberührung des Testapparates bis zum Lösen der Aufgabe) und die Anzahl der gelösten Aufgaben. Die Ergebnisse zeigen, dass Tiere aus der NE Umwelt den Testapparat (Brett) tendenziell schneller berührten. Dies war für den Rohr-Test, welcher als zweites getestet wurde, nicht mehr messbar. Die Tiere aus der NE Umwelt lösten allerdings den Rohr-Test tendenziell schneller als Artgenossen aus angereicherten Umwelten. Trotzdem fielen bei den Proportionen der gelösten Aufgaben keine Unterschiede zwischen den Haltungsumwelten auf. Testdurchgang beeinflusste die Latenzzeit bis zur Erstberührung des Testapparates dahingehend, dass die Tiere die Aufgabe beim zweiten Durchgang schneller lösten. Die Testtiere (die in der Voruntersuchung mit den unbekannten Objekten konfrontiert wurden), lösten die Brett-Aufgabe schneller als die Kontrolltiere (die in der Vorunter-suchung die Testarena leer auffanden). Im zweiten Durchgang der Rohr-Aufgabe zeigten die Testtiere, vermutlich wegen ihrer Vorerfahrung mit unbekannten Objek-ten, eine kürzere Latenzzeit bis zur Erstberührung des Testapparates als die Kon-trolltiere. Es kann geschlussfolgert werden, dass der Grad der Anreicherung in der Haltungsumwelt in dieser Studie einen tendenziellen Effekt auf die Erkundungsmoti-vation des Testapparates bei den Schweinen hatte. Die Schweine aus karger Hal-tungsumwelt lösten kognitive Aufgaben schneller. Dies wird wahrscheinlich nicht durch die höheren kognitiven Fähigkeiten dieser Tiere begründet, sondern ist eher ein erneuter Hinweis auf weniger stark befriedigtes arteigenes Explorationsverhalten („Rebound-Effekt"). Die Vorerfahrung der Testtiere mit unbekannten Objekten in der Testarena veranlasste diese wahrscheinlich auch zu einer mutigeren Herangehens-weise. Die Situation, ein Objekt in der Testarena vorzufinden, war für die Kontrolltiere eine neue Erfahrung und löste vermutlich eine vorsichtigere Herangehensweise aus, welche zu einer längeren Latenzzeit bis zur Lösung der Aufgabe führte.

Die ethologische und physiologische Komponente der Emotion ist messbar. Bei der subjektiv-erfahrenen dritten Komponente, ist dies bei Tieren nicht direkt möglich. Für Menschen dient hier die Möglichkeit des verbalen Transfers der Wiedergabe gefühlter Erfahrungen. Dies ist in Untersuchungen mit Tieren nicht möglich. Die gemessenen ethologischen und physiologischen Veränderungen bei Tieren in dieser Studie werden als positiv interpretiert. Trotzdem bleibt es unklar was die Tiere tatsächlich fühlen (Kapitel V). Die gemessenen Ergebnisse dieser Studie zeigen, dass kardiovaskuläre Parameter vielversprechende Indikatoren für positive Emotion beim Schwein sind. Allerdings sind diese Parameter leicht von anderen Einflüssen überlagert. Daher sind größere Datenmengen notwendig, um weitere Aussagen über ihre Verbindung zu ethologischen Indikatoren positiver Emotion zu treffen. Spiel und Schwanzwedel-Verhalten sowie Erkundungsverhalten sind potentielle Indikatoren positiver Emotion bei den getesteten Schweinen in dieser Studie. Ein weiterer wichtiger Aspekt in diesem Zusammenhang ist die Haltungsumwelt in welcher Schweine gehalten werden. Diese kann die Reaktionen der Tiere in Verhaltenstests stark beeinflussen. Speziell die Haltung in kargen, unangereicherten Haltungsumwelten beeinträchtigt das Erkundungsbedürfnis von Schweinen und verändert deren Reaktionen in Verhaltenstests (Kapitel V).

CHAPTER I

INTRODUCTION

INTRODUCTION

The increasing consumption of animal products over the last decades led into highly intensified animal production systems. The conditions under which animals used for this production are born, raised, kept and slaughtered are under increased public scrutiny (Kanis et al., 2003). In order to meet consumer demands, high production levels require to be maintained whilst the welfare of the animals used is ought to be improved. With the by now commonly accepted recognition that animals are sentient beings who are also able to experience emotions (Špinka, 2012), it is crucial for the successful improvement of animal welfare to develop objective methods to assess emotional states of animals (Dawkins, 2008). Emotions enable survival and therefore animals avoid situations which cause pain or fear and strive towards situations which cause pleasure and contentment. Scientists hesitated for a long time about whether non-human animals were capable of emotions, but nowadays it is accepted that almost certainly all mammals experience emotions in a comparable fashion as humans do (Špinka, 2012). Good welfare is often described as not only the absence of negative affective states but more importantly, the presence of positive affective states (Bradburn, 1969; Seligman and Csikszentmihalyi, 2000) which distinguishes a good quality of life (Morton, 2007). The study of emotions in animals developed an imbalance towards the investigation of negative emotions such as stress and fear (Boissy et al., 2007). The reason for this is probably the fact that negative affective states are expressed more intensively (as they are potentially life-threatening) whilst their positive counterparts are more subtle and harder to study (Boissy et al., 2007).

This introductory chapter aims to review the current state of knowledge on positive emotional states, their occurrence; measurability and relevance for the assessment of animal welfare (see also Table 1). In the following, firstly, positive emotions will be defined. Thereafter, the development of positive emotional states over the course of evolution and their neurobiology within the brain will be addressed. Thirdly, different ways of measuring positive emotional states will be outlined and it will be examined how the previous findings could be of use for the assessment of well-being in animals.

The present doctoral dissertation aims to identify indicators of positive emotional states. Three experiments were conducted to establish a link between behavioural indicators and physiological changes that go along with positive emotional states in pigs. In the last chapter of this thesis, the experimental results will be set into context with the findings from the scientific literature.

Definition of positive emotions

When thinking about positive emotions, the major questions arising are: How are positive emotions defined? What makes an emotion positive? And what sets a positive emotion apart from other pleasant affective states such as sensory pleasure or positive mood? What are positive emotions with regards to animals?

The study of positive emotions has received particular attention as they may serve as indicators for well-being (Diener and Seligman, 2004; Kahneman et al., 2004). A common characterization of positive emotion could very generally be described as the facilitation of approach behaviour (Cacioppo et al., 1993; Davidson, 1993). The experience of a positive emotion encourages an organism to actively interact with its environment. This connection between positive emotion and active engagement with the environment is often called "positivity offset" (Diener and Diener, 1996; Ito and Cacioppo, 1999). This positivity offset prompts individuals to approach novel objects, situations or other individuals (Frederickson and Cohn, 2008). The outline of this definition is rather general and it should be kept in mind that positive emotions also arise from various approach-avoidance situations, so can for example avoiding an unpleasant situation result in the positive emotion of relief (Frederickson and Cohn, 2008).

The common denominator of positive emotion and sensory pleasure is the accompanying pleasant feeling which is experienced subjectively. However, sensory pleasure is commonly characterized as the correction of internal trouble: for example the satiation of hunger and thirst or warming up when cold (Cabanac, 1971). In contrast, positive emotion always requires assessment of the meaning of a given stimulus. For example, food corrects the internal trouble of hunger (sensory pleasure) but can also lead to a feeling of contentment (positive emotion) (Frederickson and Cohn, 2008).

The differentiation between positive emotion and positive mood is given through the fact that a positive emotion is centred on an object, something meaningful to the individual of which it is consciously aware of. Positive emotions are typically also much more short-lived than positive moods. Positive moods in contrast, are objectless, longer lasting and the individual is only peripherally aware of this state (Oatley and Jenkins, 1996; Rosenberg, 1998).

With regards to positive emotions in animals, four categories can be defined: (i) consummatory satisfaction (e.g. contentment after eating), (ii) pleasant sensory activity (e.g. affiliation between social partners or parent and offspring, homeostatic thermal conditions such as lying in the sun or in the nest), (iii) positive expectation (e.g. anticipatory joy; Boissy et al., 2007) and (iv) emotional action activities (e.g. play or explorative behaviour; Panksepp, 2004).

Evolutionary development of positive emotions and their neurobiology in the brain

In the process of evolution, natural selection is a key mechanism. Natural selection is characterized by differences in survival and reproduction of individuals due to differences in their phenotypes (Williams, 1966). Along the process of natural selection, traits which serve an important biological function are shaped. The ability to experience various emotional states was favoured by natural selection because emotions increased an organism's ability to cope with adaptive challenges. Adaptive challenges are threats or opportunities in specific situations (occurring repeatedly over evolutionary time) for which each emotional state represents a mode of reaction both physiologically and behaviourally (Nesse, 1990). Animals therefore have evolved emotional processes to help them avoid situations which can potentially cause harm (avoid punishment) and approach situations potentially incorporating pleasure (seeking reward; Rolls, 2000; Cardinal, et al., 2002). While punishers are potentially life threatening (such as a predator attack or thermal damage), rewards are ensuring survival (i.e. food, water, shelter). Rewards and punishments are in essence the basis of emotional states (Barrett et al., 2007; Nesse and Ellsworth, 2009) and striving to the acquisition of rewards and avoiding fitness threatening punishers lies at the heart of survival (Rolls, 2005; Burgdorf and Panksepp, 2006).

It is important for an individual to be able to classify whether an external stimulus represents a reward or a punishment. Mendl and Paul (2004) describe this classification process of the stimulus as 'appraisal'. Stimuli which are classified as being pleasant, at least moderately predictable and not sudden may evoke positive emotions. It seems therefore, that specific appraisal patterns elicit corresponding emotions (Scherer, 1999). It therefore increases fitness to build sensory systems which are able to sense for example nutrient need and perform behaviours which work towards obtaining nutritious rewards when hungry.

Emotional states can be considered as interfaces between sensory inputs and action systems which create corresponding outputs (Rolls, 2000). Positive emotions are mediated through certain neural structures, the so-called reward circuits such as the mesolimbic-dopaminergic (ML-DA) axis. Once an individual encounters a potential reward (e.g. shelter, food, conspecifics), the reward circuit is activated. Within this system, neuro-modulatory substances involved in positive emotions are dopamine, opioids and oxytocin. The neurotransmitter dopamine is transmitted from the ventral tegmental area (VTA) of the brain to the nucleus accumbens (NAcc), located in the ventral striatum. Opioids (such as e.g. endorphin) released from the pituitary gland, act on the ML-DA axis either by stimulating dopaminergic VTA neurons or by increasing the concentration of dopamine in the NAcc (Mirenowicz et al., 1996: studied in monkeys; Van der Harst et al., 2003: studied in rats). The hormone oxytocin, also released from the pituitary, acts on the NAcc where it increases the release of dopamine especially in the context of social bonding (Champagne et al., 2004: studied in rats). According to the review of Berridge (1996), the positive emotional valence "wanting", regarding motivation, is mediated by the transmitter system of dopamine; whereas the emotional valence "liking", regarding pleasure, is rather mediated by the transmitter system of opioids. The transmission of dopamine therefore determines the degree of motivation for reward (differentiation between highly rewarding and less rewarding) whilst the concentration of opioids rather determine what is wanted (e.g. elicitation of feeding (animal models reviewed by Levine, 2006), social interaction or play (Vanderschuren et al., 1995: studied in rats)).

Psychostimulants such as amphetamine or cocaine elicit positive emotions in humans because the drugs activate dopamine in the ventral striatum (Drevets et al., 2001; Volkow and Swanson, 2004). Drug injections into the ventral striatum elicit vocalizations in rats which are associated with positive emotion (Burgdorf et al., 2001). The ventral striatum has also been found to be involved especially in positive emotional states such as anticipation of a reward (Knutson et al., 2001a and b: studied in humans).

Studies on the electrical stimulation of the NAcc showed that positive emotions (smiling, laughter, euphoria) could thereby be induced in humans (Heath, 1972; Okun et al., 2004). Also in non-human primates, stimulation of the NAcc elicited vocalizations which are observed to be associated with unexpectedly finding highly palatable food or being reunited with a conspecific (Jürgens, 1976). In guinea-pigs it was found, that electrical stimulation of these brain regions elicited vocalizations associated to sexual excitement and social attachment (Kyhou and Gemba, 1998).

These action systems are predominately under the control of the limbic- and further sub-neocortical systems in the brain (MacLean, 1990: reviewed in humans; Panksepp, 1998: reviewed in human and animal models). In the study of Damasio et al. (2000) human individuals were asked to think of various emotional states which they have deeply experienced themselves (e.g. happiness, sadness, anger). The tested individuals were given water infused with radioactive substances and thereafter PET images of their brains were taken. The results clearly showed increasing arousal in the sub-neocortical areas of the brain once the individuals re-lived the different emotional states and at the same time a reduction in blood flow in the neo-cortex was found. Therefore, during the experience of emotional states, a reduction of information processing in neocortical brain areas can be assumed (Liotti and Panksepp, 2004: studied in humans).

There exists only little evidence for the involvement of neocortical regions in the generation of emotional experiences. However, these regions are essential for the cognitive memories associated to various emotional states. In other words, emotional states are communicated via specific neuro-dynamic circuits to higher brain areas (orbitofrontal and medial frontal regions) so that a cognitive evaluation of the specific emotional state can take place (Alcaro et al., 2007: reviewed in human and animal models). In order to investigate neocortical involvement in play behaviour, Panksepp et al. (1994) surgically removed the neocortex of juvenile rats and observed the animal's subsequent behaviour. The authors concluded from their results that the motivation to play must be sub-cortically organized because surgically treated animals still frequently displayed play behaviour. The stimulation of sub-neocortical regions of the brain leads to much stronger emotional responses than the stimulation of neocortical regions. The sub-neocortical regions of the brain might therefore hold the emotional action systems designed to appraise objects or situations and provide behavioural reaction codes according to their survival value (Burgdorf and Panksepp, 2006: reviewed in animal and human models). However, the precise mechanisms through which behavioural patterns are established under the influence of dopamine remain unclear (Alcaro et al., 2007: reviewed in human and animal models).

Emotional responses, however, entail a cognitive component as an adaptive value (Mendl et al., 2010; Broom, 2014) as they occur tightly connected to learning and evaluation of environmental stimuli (Broom, 2014). An example for emotional involvement in cognition is the phenomenon of cognitive bias. An individual's interpretation of an ambiguous situation is influenced by its underlying emotional state. Individuals in a negative emotional state interpret the ambiguous stimulus as negative and react to it as if a negative outcome will occur and vice versa (Mendl et al., 2009; Broom, 2014). Emotional state is also affecting learning behaviour. In an experiment of Carey and Fry (1993), pigs were trained to show an operant response when injected with an anxiolytic drug and a different operant response when injected with saline. When the pigs were then confronted with other anxiety-inducing stimuli (transportation, novel object, novel pen or an unfamiliar conspecific), they showed the same operant response as if injected with the anxiolytic drug. Thus, the operant response they chose to display indicated their emotional state of anxiety.

Measurement of positive emotions

For the study of positive emotions, different approaches have been applied including physiological and behavioural parameters. In physiological studies, one major focus is on the neurobiological functions. With the help of modern brain imaging, changes within the brain accompanying positive emotional states can be mapped. One technique which is frequently used in this context is functional magnetic resonance imaging (fMRI). Here, changes in brain activity are detected through measuring changes in blood flow (Huettel et al., 2009). fMRI brain imaging is suitable for detecting the cognitive processes associated to an emotional state as they are generated very rapidly in response to sensual inputs, e.g. memory of an event which triggered an emotional state in the past. However, the actual emotional state of an organism emerges less rapidly and is therefore suggested to map core affective states in the brain with an imaging technique such a positron emission tomography scanning (PET scan; Burgdorf and Panksepp, 2006). This imaging technique is based upon radioactive tracers with which the test subjects are injected. The tracer substance emits gamma rays which are detected by the tomograph. The concentration of the tracer substance indicates the amount of metabolic activity within the specific brain regions (Bailey et al., 2005). Increases in activity of specific brain regions could therefore indicate involvement of said region in the generation and experience of various emotional states.

As mentioned above, the limbic system, located in sub-neocortical areas of the brain, is largely involved in generating emotions. A main efferent pathway of the limbic system is the autonomic nervous system (ANS) consisting of sympathetic and parasympathetic control mechanisms which reflect the physiological state of an organism, e.g. stress or homeostasis (Boissy et al., 2007). This physiological state is generally reflected by heart rate and heart rate variability (HVR) as the heart is both under the control of the sympathetic and parasympathetic branch of the ANS. Through direct innervation of the heart through sympathetic and parasympathetic fibres, emotions therefore have a large effect on cardiac activity (Saul, 1990). The pulses of heart beat are generated by the sinoartrial node (SN), who acts as primary pacemaker. Through depolarisation of cells in the SN, electrical stimulation then activates the heart muscles. The SN is innervated by sympathetic and parasympathetic nerve fibres which, depending on which fibres dominate, influence the heart rate and HRV in distinct ways (Von Borell et al., 2007). Sympathetic control of heart activity is associated with an increase of heart beats and decrease of heart rate variability, commonly associated with stress responses and negative emotional states whereas parasympathetic control of cardiac activity is associated with lower heart rate and higher HRV, commonly associated with relaxation and positive emotional states (Rainville et al., 2006; Von Borrell et al., 2007). HRV therefore provides a potential indication on the emotional state experienced by an organism. However, a rise in HR can not only be the result of a dominant sympathetic activation, but also that of a decrease in vagal activation or from changes in both nervous systems (Von Borell et al., 2007). Also, HR can be influenced by neurotransmitters such as Acetylcholin (decreasing HR) or adrenalin and cortisol (increasing HR). In a study by McCraty et al. (1995), human test persons were asked to mentally recall and visualize past positive experiences. Electrocardiographic measurements were then taken and it was found that test persons' HRV increased in response to the positive emotional state. The authors also report a lowering of blood pressure when patients experience a positive emotional state (McCraty et al., 1995). Fox (1989) described a close relation between HVR and their reactivity to positive or negative events in human infants. Children with high HRV were more approachful and more inclined to engage with their mothers and strangers at play (Fox, 1989). HR and HRV are therefore insightful indicators for measuring emotional changes physiologically. Measurement of HR and HRV is mostly carried out non-invasively and handling of the required equipment is fairly straight forward.

A further physiological indicator for positive emotion was found in studies assessing saliva samples of human test subjects. Test subjects experiencing a positive emotional state showed increases in the saliva concentration of 20-200 kD proteins whereas subjects experiencing negative emotions had a decrease of said proteins (Grigoriev et al., 2003). Interestingly, 20-200kD proteins are associated with α-amylase and increases of these proteins are related to relaxation-induced parasympathetic secretion of saliva through the parotid and submaxillary gland as found by Morse et al., 1989. In this study, human adult patients' anxiety levels were investigated based on a questionnaire and saliva samples taken before and after a dental procedure. With half of the test persons, relaxation techniques (meditation and hypnosis) were practiced in combination with receiving a sedative prior to the dental procedure. The other half of the patients were only given a sedative without practising relaxation techniques. All persons of the test group that received both the relaxation technique and the sedative reported reduced anxiety levels. Whereas in the test group that only received the sedative (and not the relaxation technique), reduced anxiety was only found in one third of the group (Morse et al., 1989). According to Boissy et al. (2007) the relation between salivary amylase and emotional states remains unclear but further insights into such measures in animals could be a useful tool to monitor their emotional states.

Emotional limbic activity in the brain was also found to influence immunological processes with immunoglobulin A (IgA) playing a crucial role in the immune function (Haas and Schauenstein, 1997). There are studies correlating immune activation with anxious or depressive behaviours in humans (Dantzer, 2001; Dunn et al., 2005). Salivary IgA concentration was found to be modulated by experiencing positive emotions (McCraty et al., 1996; Watanuki and Kim, 2005). A study of Stone et al.

(1987) investigated positive versus negative mood of male students. Test persons were asked to fill out questionnaires describing their mood as well as taking salivary samples three times weekly. The authors found, that the saliva IgA levels were depressed on days of high negative mood and elevated on days with high positive mood. Further, daily stress increased negative mood and decreased positive mood along with a reduction of IgA. The authors concluded that stress increases the likelihood of encountered viruses gaining entry into the body (Stone et al., 1987). Salivary composition could therefore potentially be used as a monitoring tool for emotional states. However, immune parameters such as IgA are under the influence of multidimensional factors and can easily be biased by unapparent infections in the body and therefore the interpretation of changes in their concentration can be ambiguous (Boissy et al., 2007).

For the measurement and assessment of positive emotion, the insight into physiological and neurobiological pathways provides very important understanding of the underlying mechanisms. However, physiological changes are merely action patterns dictating an individual how best to behave in a specific situation. Behavioural changes can therefore also be used as a reliable measure to assess emotionality in animals. Unconditioned, spontaneous behaviour of an individual towards a situation or object can be regarded as action patterns which reflect the individuals' accompanying emotional state. Generally, approach or avoidance can be cues to the valence an object or situation might have for the individual (Paul et al., 2005). In the case of a positive underlying emotional state and thus a positive appraisal of a situation or object, animals' behaviour might be of exploratory, playful or consumptive nature (Špinka et al., 2001; Paul et al., 2005). In contrast to spontaneously occurring behaviour, behaviour tests can also be used to assess animal emotion. In a recent study of Reimert et al. (2013), it was found that various behaviours tended to occur significantly more often when the test animals (pigs) were given a positive treatment (two acquainted pigs were given access to a large area filled with peat, straw and chocolate raisins), thus probably facilitating positive emotion compared to the negative treatment (isolation from conspecific in small barren pen). It was found that the test animals showed significantly more play behaviour, more tail wagging and emitted more barking vocalizations when given the positive treatment. In contrast, when exposed to the negative treatment, the test animals exhibited more standing alert behaviour, made more escape attempts, urinated more, held their ears in a more backwards position and emitted more high pitched vocalizations (Reimert et al., 2013).

Relevance of positive emotions for the assessment of animal welfare

The understanding of animal emotion is crucial because the mere existence of these emotional states is the underlying factor of growing concerns of both public and scientific interest in animal welfare. In a recent review on the study of emotions in pigs, Murphy et al., (2014) state that experiences of positive emotion contribute to good welfare of an animal. However, in modern pig production systems the basic needs of pigs regarding cognitive challenge and the ability to express species-specific behaviour are not met. As a consequence, pigs kept in such production systems are predominately experiencing negative emotional states and their welfare is therefore reduced (Murphy et al., 2014). It is important to be able to assess under which circumstances animals experience positive emotional states (Boissy et al., 2007), and facilitate these circumstances in order to improve the animals' quality of life.

The experience of positive emotions can be facilitated by the provision of rewards which the animal finds desirable. Spruijt et al. (2001) suggested using anticipation behaviour for the study of emotional states. During Pavlovian conditioning, behaviours between the presentation of the cue predicting the reward and the actual presentation of the reward can be suitable indicators. Spruijt et al. (2001) concluded that the anticipation of a positive event or stimulus can induce a positive emotional state in the animal. Dudink et al. (2006) examined the effects of anticipation in piglets at weaning. They observed that the anticipation of enrichments resulted in major behavioural changes of the piglets, even more so, the anticipation resulted in more profound effects than the presentation of the enrichment alone. In anticipation of access to enrichment the piglets showed more play behaviour and less aggressive behaviour after weaning. Resulting from these behavioural changes, the piglets also had a decreased amount of bodily injuries after weaning (Dudink et al., 2006).

Another factor contributing to the well-being of animals in captivity is their ability to control and cope with the environment they live in. Deliberate actions the animal can take upon facing a challenge within its environment which enable a positive outcome facilitate controllability and therefore positive emotion (Boissy et al., 2007). In this context, cognitive enrichment could offer an interesting approach. Puppe et al. (2007) designed a food rewarded learning system in which pigs were conditioned to associate an individual call type emitted from the feeding station to the arrival of a food reward. All animals learned to discriminate calls which announced food rewards and were also willing to work for food by pressing a button. Further, the conditioned pigs showed less fearful and anxious behaviour in an open-field test compared to a conventionally fed control group. The authors concluded that the cognitive enrichment could have induced frequent positive emotional states (Puppe et al., 2007).

There are numerous attempts to add stimuli to otherwise barren living environments in order to reduce indicators of poor welfare (e.g. chains or wooden bars for chewing), however these strategies rather try to reduce indicators of poor welfare instead of increasing indicators of good welfare (e.g. play). Studies which enabled young animals to have more space showed that the surplus of space facilitated an increase in play behaviour both in calves (Jensen et al., 2000) and in pigs (Blackshaw et al., 1997).

Aim and outline of the thesis

Improvement of animal welfare lies at the heart of animal welfare science, especially regarding the large numbers of livestock animals intensively bred and kept for the supply of animal products for rising human consumption. As intensive livestock production is unlikely to reduce with ever increasing world population in the near future, studies on animal welfare are crucial for the improvement and solution to animal welfare problems for livestock animals. Animal welfare problems such as abnormal behaviour or deficient health are mere red flags of underlying difficulties which go along with "housing" animals and thereby reducing the scope of their natural behaviour. Deeper understanding of the way livestock animals appraise their surroundings in intensive husbandry systems may enable us to determine what causes negative experiences but also what causes positive experiences. Especially the experience of positive emotional states may increase animal welfare greatly and may also weigh against unavoidable husbandry practices that cause negative experiences in animals. If an animal has the opportunity to frequently experience positive emotional states, it leads a life that is of value to the individual itself (McMillan, 2005).

The main aim of the present thesis was to investigate physiological and behavioural indicators of positive emotion in pigs. Research into emotions of animals is a growing interest from the late 20th century onward. The assessment of pain and suffering has been a focus in the endeavour to improve animal welfare and various methods have been developed to that effect by animal welfare scientists (Boissy et al., 2007). Despite the fact that assessment of negative emotional states is of high importance, as they are potentially life-threatening, investigation into the assessment of positive experiences is not as distinctly developed and often characterized by disagreements. Still, the experience of positive emotional states is as crucial to good animal welfare as the avoidance, or manageability, of negative emotions. Therefore, it is a central issue to be able to identify the circumstances under which animals experience positive emotion (Boissy, et al., 2007). The interplay of three components making up an emotion (behaviour, physiology and consciousness) is highly complex. It remains unclear how positive emotions are expressed behaviourally, also taking into account

27

variations between and within species, how this behaviour is regulated physiologically and, lastly, how the animal is consciously experiencing positive emotion.

The main aim of the present thesis is to investigate potential behavioural indicators of positive emotion in young pigs in combination with underlying physiological markers. Three studies were conducted:

- It was hypothesized that objects and materials of more complex nature would elicit more exploratory behaviour and an increase in parasympathetic influence in the cardiovascular system of pigs. Male and female juvenile pigs were therefore confronted with stimuli which potentially trigger a positive emotional state (i.e. given access to rooting material, a wallow or hidden food items). When the animals had access to the different materials and objects, video recordings were used to analyse their behavioural expression, especially the exploratory behaviour of the animals. Their corresponding heart rate and heart rate variability during testing was measured and analysed (chapter 2).
- Pigs housed in stimulus and substrate enriched environments could display species specific behaviour more frequently and it is assumed that they experience positive affective states more frequently. We hypothesized that the enrichment level in the home environment would also affect their behaviour in a test situation. The influence of varying levels of enrichment in housing environments and the effect of novelty on vascular parameters (HR, HRV), play and tail wagging behaviour was therefore studied in 36 female juvenile pigs during novel object tests (chapter 3). Play and tail wagging behaviour were measured as possible determinants of positive emotion and were merged as Positive Emotion Score (PES).
- We hypothesized, that pigs reared in an environment containing more stimuli would show more cognitive flexibility which would enable them to solve a cognitive task faster than conspecifics housed in an impoverished environment. The ability of 36 female growing pigs to solve a cognitive task was tested against the background of varying enrichment levels in housing environment and novelty of behavioural testing. The latency to touch the test apparatus and the latency to solve the task was assessed and compared between treatments and environments.

Table 1: Summary of selected studies on positive emotions in animals

Theoretical approach	Species, N, Age, Gender	Experimental design/ treatment	Parameters recorded	Results	Publication
Anticipation	Chickens, N = 12, aged 14 weeks, female	Trace conditioning regime: Conditioned stimulus (CS, light cue) paired with unconditioned stimulus (US, mealworm). Time between presentation of CS and US delayed. Control animals were subjected to CS and US in random order.	Anticipatory, consummatory behaviour	Conditioned hens reacted to CS with increased levels of anticipatory behaviour. No difference was found with regards to consummatory behaviours compared to controls.	Moe et al. (2009)
Anticipation, behavioural and physiological indicators of emotion	Sheep, N = 45, aged 5 months, female	Operant conditioning task to obtain small or large food reward. Half of the animals were shifted to the large or the small reward, other half remained at the same reward amount. Then, shifted animals were backshifted to their initial reward amount and other previously remaining animals were subjected to a complete reward distinction.	Behaviour, cardiac activity	Reward decrease and reward distinction: more locomotor activity, operant task was performed at higher frequency but less efficient, parasympathetic influence of HR decreased. Reward increase: decrease in attempts to perform operant task. Animals form expectations and discrepancy from these expectations influences emotional response.	Greiveldinger et al. (2011)

Table 1 continued

Anticipation, behavioural indicators of emotion	Pigs, N = 24, aged 12weeks, male	Half of the animals trained to anticipate pleasurable or aversive event. Naïve conspecifics observed for emotional contagion during anticipation together with trained pigs.	Behavioural indicators of positive and negative emotional state.	Anticipation of pleasurable event: play, barks, tail wagging. Negative event: freezing, defecating, urinating, escape attempts, high-pitched vocalization, tail low, ears back. Emotional contagion indicated through behaviours of trained and naïve pigs.	Reimert et al. (2013)
Behavioural indicators of emotion	Cows, N = 20, female	Experimental situation: 1. Inducing frustration (access to food is denied) 2. Inducing positive affect (access to food is allowed). Half of the cows in experimental situation 1 were given an anxiety-reducing drug (diazepam).	Percentage of eye white as an indicator of emotion	Calculated percentage of eye white was larger in frustrated cows. Treatment with diazepam reduced the percentage of eye-white of frustrated cows. Eye – white served as a reliable emotional indicator.	Sandem et al. (2006)

30

Table 1 continued

Behavioural, physiological indicators of emotion	Sheep, N = 19, aged 4 months, female	Experimental situations (each lasting 4 min): 1. separation from group members (negative valence) 2. standing in the feeding area (intermediate valence) 3. being voluntarily groomed by a familiar human (positive valence)	Behavioural reaction (ear posture, relative eye aperture) and physiological reaction (cardiovascular parameters, respiration, body surface humidity, temperature) to situations of varying emotional valence	Sheep in situation 3 (positive valence): Few ear posture changes, few forward ear postures, low eye aperture, low variance of body surface humidity (values linearly inclined in situation 2 and 1), long inter-beat intervals and high HRV (values linearly declined in situation 2 and 1).	Reefman et al. (2009)
Behavioural indicators of emotion	Cows, N = 13, female	Experimental animals were stroked on head, neck and withers for 5 min.	Ear postures, ear posture changes	During stroking, cows held their ears in a hanging position, perpendicular to the head. Ear posture changes were more pronounced during stroking compared to pre-stroking and post-stroking phases. Relaxed ear postures were associated with positive, low arousal emotional state.	Proctor and Carder (2014)

Table 1 continued

Cognitive Bias Test	Pigs, N=10, aged 12 weeks, male	Half housed in enriched environment, other half in barren environment.	Approach Behaviour, Latency to approach	Animals housed in enriched environment were more likely and faster to approach the hatch in response to the ambiguous cue. Indicative of a positive affective state, the animals housed in enriched pens showed a more optimistic judgement bias.	Douglas et al. (2012)
Cognitive Enrichment	Pigs, N = 56, aged 7 weeks, male	Learning behaviour in call-feeding-station (classical and operant conditioning using acoustic cues) for experimental animals. Control group was fed conventionally. Behaviour of test and control group was then compared in home environment and open field test / novel object test.	Discrimination ability, behaviour	Experimental animals: successfully discriminated acoustic cue, were willing to work for food (push button), showed more locomotor behaviour and less belly nosing in the home environment, reduced excitement and fear behaviour in the test situation than the control pigs. Behavioural differences were more pronounced the longer the experimental animals were confronted with the cognitive challenge.	Puppe et al. (2007)

Table 1 continued

| Cognitive Enrichment | Pigs, N = 48, aged 10 weeks, male | Effect of cognitive enrichment (food reward upon correct discrimination of an acoustic cue and pressing of a button) of experimental animals on behaviour and vascular parameters in the home pen and an open field test compared to conventionally fed control animals. | HR, HRV and behaviour during feeding and in OFT | Upon feeding announcement, both experimental and control animals showed high HR and low HRV. During feeding, experimental animals showed reduced HR and increased HRV while control animals' HR remained very high (higher number of agonistic interactions during feeding was observed in control animals). During open-field-test, experimental animals showed more explorative behaviour and less fearful behaviour than control animals. Cognitive enrichment led to a more relaxed feeding situation and longer lasting positive emotions. | Zebunke et al., (2013) |

Table 1 continued

Emotional contagion	Pigs, N = 96, aged 9 weeks, male	Behavioural indicators of emotion were observed after experiencing pleasurable or aversive event for treated pigs and naïve pen-mates that did not experience events themselves.	Ethogram recording behaviours indicative for positive and negative affective state.	Treated pigs showed behaviours indicative of negative affective state after the aversive event. Also naïve pigs adopted these negative behaviours. Carryover effects of the pleasurable event were indicated by increased body contact both in treated and naïve pigs. Also naïve pen mates were emotionally affected. Negative and positive events may have carryover effects.	Reimert et al. (2017)
Environ-mental Enrichment	Pigs, N = 307, birth to weaning at 28 days of age, female and male	Three different farrowing environments: a) Standard parallel crate (0.8 x 2.1m) b) Round crate giving the sow more space (1.85m diameter) c) Open farrowing pen (2.1 x 2.1m)	Play Behaviour	Individual play occurred most in the round crate, object play (pushing, biting, sniffing inanimate objects) occurred most in the standard crate. Social play: Nudging occurred mostly in the open pen, pushing occurred mainly in the standard crate. Aggression (play ending in fighting): mostly in the standard crate (27.8%) and less in the round crate (13.4%) or the open pen (12.5%).	Blackshaw et al. (1997)

Table 1 continued

Environmental Enrichment	Rats, N = 60, aged 7-8 weeks, male	Standard and Enriched housing of test groups (conditioning regime: sound cue + sucrose reward) and control groups (only sound cue).	Anticipatory Behaviour shown between sound cue and sucrose reward.	Anticipatory response to the sound cue was more pronounced in rats living in standard environment indicating a higher sensitivity to rewards than enriched animals. This might come from stress caused by the disability to satisfy behavioural needs in impoverished conditions.	Van der Harst et al. (2003)
Open Field test (OFT)	a. Pigs, N = 24, aged 5 weeks, 13 males, 11 females b. Pigs, N = 20, aged 5 weeks, 10 males, 10 females c. Pigs, N = 12, aged 5 weeks, 6 males, 6 females	a. OFT with and without Azaperone (stress-reducing drug). Saline control. b. OFT with and without conspecific c. OFT on two consecutive days	Explorative behaviour, vocalization	a. Azaperone treated pigs spent more time exploring, were more active and vocalized less than saline controls. b. Pigs experiencing OFT with companion vocalized less and explored more. c. Pigs experiencing OFT for the second time were less active, vocalized less and showed less explorative behaviour than the first time. PCA showed that the manipulations or previous experience are reflective of underlying emotionality.	Donald et al. (2011)

35

References

Alcaro, A., Huber, R., Panksepp, J. 2007. Behavioral functions of the mesolimbic dopaminergic system: An affective neuroethological perspective. Brain Res. Rev. 56, 283-321.

Bailey, D.L., Townsend, D.W., Valk, P.E., Maisey, M.N. 2005. Positron Emission Tomography: Basic Sciences. Secaucus, NJ: Springer-Verlag

Barrett L. F., Mesquita B., Ochsner K. N., Gross J. J. 2007. The experience of emotion. Ann. Rev. Psychol. 58, 373–403.

Berridge K.C. 1996. Food reward: brain substrates of wanting and liking. Neuroscience and Biobehav. Rev. 20, 1–25.

Blackshaw J.K., Swain A.J., Blackshaw A.W., Thomas F.J.M., Gillies K.J. 1997. The development of playful behavior in piglets from birth to weaning in three farrowing environments. Appl. Anim. Behav. Sci. 55, 37–49.

Boissy, A., Manteuffel, G., Jensen, M.B., Oppermann Moe, R., Spruijt, B., Keeling, L.J., Winckler, C., Forkman, B., Dimitrov, I., Langbein, J., Bakken, M., Veissier, I., Aubert, A. 2007. Assessment of positive emotions in animals to improve their welfare. Physiol. Behav. 92, 375-397.

Bradburn N.M. 1969. The structure of psychological well-being. Chicago: Aldline Publishing Company

Broom, D. M. 2014. Sentience and animal welfare. CABI, Wallingford, UK.

Burgdorf, J., Knutson, B., Panksepp, J., Ikemoto, S. 2001a. Nucleus accumbens amphetamine microinjections unconditionally elicit 50-kHz ultrasonic vocalizations in rats. Behav. Neurosci. 115, 940–944.

Burgdorf, J., Panksepp, J. 2006. The neurobiology of positive emotions. Neuroscience and Biobehav. Rev. 30, 173-187.

Cabanac, M. 1971. Physiological role of pleasure. Science 173, 1103–1107.

Cacioppo, J.T., Priester, J.R., Berntson, G.G. 1993. Rudimentary determinants of attitudes: II. Arm flexion and extension have differential effects on attitudes. J. Pers. Soc. Psychol. 65, 5-17.

Cardinal, R.N., Parkinson, J.A., Hall, J., Everitt, B.J. 2002. Emotion and motivation: the role of the amygdala, ventral striatum and prefrontal cortex. Neurosci. Biobehav. Rev. 26, 321-352.

Carey, M. P., Fry, J. P. 1995. Evaluation of animal welfare by the self-expression of an anxiety state. Lab. Anim. 29, 370-379.

Champagne F.A., Chretien P., Stevenson C.W., Zhang T.Y., Gratton A., Meaney M.J. 2004. Variations in nucleus accumbens dopamine associated with individual differences in maternal behavior in the rat. J. Neurosci., 4113– 4123.

Damasio, A.R., Grabowski, T.J., Bechara, A., Damasio, H., Ponto, L.L., Parvizi, J., Hichwa, R.D., 2000. Sub-neocortical and cortical brain activity during the feeling of self-generated emotions. Nat. Neurosci. 3, 1049–1056.

Dantzer R. 2001. Cytokine-induced sickness behavior: where do we stand? Brain Behav. Immun. 15, 7–24.

Davidson, R.J. 1993. The neuropsychology of emotion and affective style. In: Lewis, M., Haviland, J.M. Handbook of Emotions. p. 143-154. New York: Guilford Press

Dawkins, M.S. 1990. From an animals' point of view: Motivation, fitness and animal welfare. Behav. Brain Sci. 13, 1-61.

De Waal, F. 2011. What is an animal emotion? Annals of the New York Academy of Sciences 1224, 191-206.

Diener, E., Diener, C. 1996. Most people are happy. Psychol. Sci. 7, 181–185.

Diener, E., Seligman, M.E.P. 2004. Beyond money: Toward an economy of well-being. Psychol. Sci. Public Interest 5, p. 1–31.

Donald, R.D., Healy, S.D., Lawrence, A.B., Rutherford, K.M.D. 2011. Emotionality in growing pigs: Is the open field a valid test? Physiolog. Behav. 104, 906-913.

Douglas, C., Bateson, M., Walsh, C., Bédué, A., Edwards, S.A. 2012. Environmental Enrichment induces optimistic cognitive biases in pigs. Appl. Anim. Behav. Sci. 139, 65 – 73.

Drevets, W., Gautier, C., Price, J., Kupfer, D., Kinahan, P., Grace, A. 2001. Amphetamine-induced dopamine release in human ventral striatum correlates with euphoria. Biolog. Psychiatry 49, 81–96.

Dudink, S., Simonse, H., Marks, I., De Jonge, F., Spruijt, B. 2006. Announcing the arrival of enrichment increases play behaviour and reduces weaning-stress-induced behaviours of piglets directly after weaning. Appl. Anim. Behav. Sci. 101, 86-101.

Dunn A.J., Swiergiel A.H., de Beaurepaire R. 2005. Cytokines as mediators of depression: what can we learn from animal studies? Neurosci. Biobehav. Rev. 29, 891–909.

Fox, N.A. 1989. Psychophysiological correlates of emotional reactivity during the first year of life. Dev. Psychol. 25,364-372.

Frederickson, B.L., Cohn, M.A. 2008. Positive Emotions. p. 777-797 In: Lewis, M., Haviland-Jones, J.M., Feldmann-Barrett, L. (3rd Ed) Handbook of Emotions. The Guilford Press.

Greiveldinger, L., Veissier, I., Boissy, A. 2011. The ability of lambs to form expectations and the emotional consequences of a discrepancy from their expectations. Psychoneuroendocrinology 36, 806 – 815.

Grigoriev, I. V., Nikolaeva, L. V., & Artamonov, I. D. 2003. Protein content of human saliva in various psycho-emotional states. Biochem. (Mosc.), 68, 405-406.

Haas H.S., Schauenstein K. 1997. Neuroimmunomodulation via limbic structures—the neuroanatomy of psychoimmunology. Progr. Neurobiol. 51, 195–222.

Heath, R.G. 1972. Pleasure and brain activity in man. Deep and surface electroencephalograms during orgasm. J. Nerv. Ment. Dis. 154, 3–18.

Huettel, S.A., Song, A.W., McCarthy, G. 2009. Functional Magnetic Resonance Imaging (2.ed.), Massachusetts: Sinauer

Jensen M.B., Kyhn R. 2000. Play behavior in group housed dairy calves, the effect of space allowance. Appl. Anim. Behav. Sci. 67, 35–46.

Jürgens, U. 1976. Reinforcing concomitants of electrically elicited vocalizations. Exp. Brain Res. 26, 203–214.

Kahneman, D., Kreuger, A. B., Schkade, D. A. 2004. A survey method for characterizing daily life experience: The day reconstruction method. Science 306, 1776–1780.

Kanis, E., Groen, B. F., De Greef, K.H. 2003. Societal concerns about pork and pork production and their relationships to the production system. J. Agric. Environ. Ethics 16, 137-162.

Knutson, B., Adams, C., Fong, G., Hommer, D., 2001a. Anticipation of monetary reward selectively recruits nucleus accumbens. J. Neurosci. 21, 1–5.

Knutson, B., Fong, G.W., Adams, C.M., Varner, J.L., Hommer, D. 2001. Dissociation of reward anticipation and outcome with event-related fMRI. Neuroreport 12, 3683–3687.

Kyuhou, S., Gemba, H. 1998. Two vocalization-related subregions in the midbrain periaqueductal gray of the guinea pig. Neuroreport 9, 1607–1610.

Levine A.S. 2006. The animal model in food intake regulation: examples from the opioid literature. Physiol. Behav. 89, 92–96.

Liotti, M., Panksepp, J., 2004. Imaging human emotions and affective feelings, Implications for biological psychiatry. In: Panksepp, J., Textbook of Biological Psychiatry. Wiley, Hoboken, NJ, p. 33–74.

Ito, T.A., Cacioppo, J.T. 1999. The psychophysiology of utility appraisals. In: Kahneman, D., Diener, E., Schwartz, N. Well-being: Foundations of hedonic psychology. New York: Russell Sage Foundation. 470–488.

Maier S.F., Watkins L.R. 1998. Cytokines for psychologists: implications of bidirectional immune-to-brain communication for understanding behavior, mood, and cognition. Psychol. Rev. 105, 83–107.

McCraty, R., Atkinson, M., Tiller, W.A., Rein, G., Watkins, A.D. 1995. The effects of emotions on short-term power spectrum analysis of heart rate variability. Am. J. Cardiol. 76, 1089-1093.

McCraty, R., Atkinson, M., Rein, G., Watkins, A. D. 1996. Music enhances the effect of positive emotional states on Salivary IgA. Stress Med. 12, 167–175.

MacLean, P.D. 1990. The Triune Brain in Evolution. Plenum Press, New York.

McMillan, F.D., 2005. The concept of quality of life in animals. In: McMillan, F.D. (Ed.), Mental Health and Well-Being in Animals. Blackwell Publishing, Ames, IO, p. 183–200.

Mendl, M., Paul, E.S. 2004. Consciousness, emotion and animal welfare: insights from cognitive science. Anim. Welf. 13, 17-25.

Mendl, M., Burman, O. H., Parker, R. M., Paul, E. S. 2009. Cognitive bias as an indicator of animal emotion and welfare: emerging evidence and underlying mechanisms. Appl. Anim. Behav. Sci. 118, 161-181.

Mendl, M., Burman, O. H., Paul, E. S. 2010. An integrative and functional framework for the study of animal emotion and mood. Proc. R. Soc. Lond. B: Biol. Sci. 277, 2895-2904.

Mirenowicz J., Schultz W. 1996. Preferential activation of midbrain dopamine neurons by appetitive rather than aversive stimuli. Nature 379, 449–51.

Moe Oppermann, R., Nordgreen, J., Janczak, A.M., Spruijt, B.M., Zanella, A.J., Bakken, M. 2009. Trace classical conditioning as an approach to the study of reward-related behaviour in laying hens: A methodological study. Appl. Anim. Behav. Sci. 121, 171 – 178.

Morse, D.R., Schacterle, G.R., Furst, M.L., Bose, K. 1981. Stress, relaxation and saliva: A pilot study involving endodontic patients. Oral Surgery, Oral Medicine, Oral Pathol. 52, 308-313.

Morton D.B. 2007. A hypothetical strategy for the objective evaluation of animal well being and quality of life using a dog model. Anim. Welf. 16, 75-81.

Murphy, E., Nordquist, R.E., Van der Straay, F.J. 2014. A review of methods to study emotion and mood in pigs, Sus scrofa. Appl. Anim. Behav. Sci. 159, 9-28.

Nesse, R.M. 1990. Evolutionary Explanations of Emotions. Hum. Nat. 1, 261-289.

Nesse R. M., Ellsworth P. C. 2009. Evolution, emotions, and emotional disorders. Am. Psychol. 64, 129–139.

Oatley, K., Jenkins, J. M. 1996. Understanding emotions. Cambridge, MA: Blackwell.

Okun, M.S., Bowers, D., Springer, U., Shapira, N.A., Malone, D., Rezai, A. R., Nuttin, B., Heilman, K.M., Morecraft, R.J., Rasmussen, S.A., Greenberg, B.D., Foote, K.D., Goodman, W.K. 2004. What's in a 'smile?' Intra-operative observations of contralateral smiles induced by deep brain stimulation. Neurocase 10, 271–279.

Panksepp, J. 1998. Affective Neuroscience, The Foundations of Human and Animal Emotion. Oxford University Press, New York.

Panksepp, J. 2004. Textbook of Biological Psychiatry. Wiley, Hoboken, NJ.

Paul, E.S., Harding, E.J., Mendl, M. 2005. Measuring emotional processes in animals: the utility of a cognitive approach. Neurosci. Biobehav. Rev. 29. 469-491.

Puppe, B., Ernst, K., Schön, P.C., Manteuffel, G. 2007. Cognitive enrichment affects behavioural reactivity in domestic pigs. Appl. Anim. Behav. Sci. 105, 75-86.

Proctor, H.S., Carder, G. 2014. Can ear postures reliably measure the positive emotional state of cows? Appl. Anim. Behav. Sci. 161, 20-27.

Rainville P., Bechara A., Naqvi N., Damasio A.R., 2006. Basic emotions are associated with distinct patterns of cardiorespiratory activity. Internat. J. Psychophysiol. 61, 5–18.

Reefmann, N., Wechsler, B., Gygax, L. 2009. Behavioural and physiological assessment of positive and negative emotion in sheep. Anim. Behav. 78, 651-659.

Reimert, I., Bolhuis, E.J., Kemp, B., Rodenburg, B. 2013. Emotions on the loose: emotional contagion and the role of oxytocin in pigs. Anim. Cogn. 18, 517-532.

Reimert, I., Fong, S., Rodenburg, B.T., Bolhuis, J.E. 2017. Emotional states and emotional contagion in pigs after exposure to a positive and negative treatment. Appl. Anim. Behav. Sci. 193, 37-42.

Rolls, E.T. 2000. The brain and emotion. Behav. Brain Sci. 23, 177-234.

Rolls E. T. 2005. Emotion explained. Oxford, UK: Oxford University Press.

Rosenberg, E. L. 1998. Levels of analysis and the organization of affect. Rev. Gen. Psychol. 2, 247–270.

Sandem, A.I., Janczak, A.M., Braastad, B.O. 2006. The use of diazepam as a pharmacological validation of eye white as an indicator of emotional state in dairy cows. Appl. Anim. Behav. Sci. 96, 177-183.

Saul, J.P. 1990. Beat-to-beat variation of heart rate reflects modulation of cardiac autonomic outflow. Physiol. 5, 32–7.

Scherer, K.R. 1999. Appraisal Theories. In: Dalgleish,T., Power, M. Handbook of Cognition and Emotion, p. 637-663. John Wiley and Sons: Chichester, UK

Seligman M.E.P., Csikszentmihalyi M. 2000. Positive psychology: an introduction. Am. Psychol. 55, 5-14.

Špinka, M. 2012. Social dimensions of emotions and its implication for animal welfare. Appl. Anim. Behav. Sci. 138, 170-181.

Špinka, M., Newburry, R.C., Bekoff, M. 2001. Mammalian play: training for the unexpected. Q. Rev. Biol. 76, 141-168.

Spruijt B.M., van den Bos R., Pijlman F.T. 2001. A concept of welfare based on reward evaluating mechanisms in the brain: anticipatory behavior as an indicator for the state of reward systems. Appl. Anim. Behav. Sci. 72, 145–171.

Stone, A,A., Cox, D.S., Valdimarsdottir, H., Jandorf, L., Neale, J.M. 1987. Evidence That Secretory IgA Antibody Is Associated With Daily Mood. J. Pers. Soc. Psychol. 52, 988 – 993.

Von Borell, E., Langbein, J., Després, G., Hansen, S., Leterrier, C., Marchant-Forde, J., Marchant-Forde, R., Minero, M., Mohr, E., Prunier, A., Valance, D., Veissier, I. 2007. Heart rate variability as a measure of autonomic regulation of cardiac activity for assessing stress and welfare in farm animals – A review. Physiol. Behav. 92, 293-316.

Van der Harst J.E., Baars A.M., Spruijt B.M. 2003. Standard housed rats are more sensitive to rewards than enriched housed rats as reflected by their anticipatory behavior. Behav. Brain Res. 142, 151-6.

Vanderschuren L.J., Spruijt B.M., Hol T., Niesink R.J.M., van Ree J.M. 1995. Sequential analysis of social play behavior in juvenile rats: effects of morphine. Behav. Brain Res. 72, 89–95.

Volkow, N.D., Swanson, J.M. 2003. Variables that affect the clinical use and abuse of methylphenidate in the treatment of ADHD. Am. J. Psych. 160, 1909–1918.

Watanuki S., Kim Y.K. 2005. Physiological responses induced by pleasant stimuli. J. Physiol. Anthropol. Appl. Human Sci. 24, 135-8.

Williams, G.C. 1966. Adaptation and natural selection: A critique of some current evolutionary thought. Princeton: Princeton University Press

Zebunke, M. 2009. Dissertation: Affektive und emotionale Reaktionen von Schweinen im Kontext von kognitiver Umweltanreicherung. Universität Rostock.

Zebunke, M., Puppe, B., Langbein, J. 2013. Effects of cognitive enrichment on behavioural and physiological reactions of pigs. Physiol. Behav. 118, 70-79.

CHAPTER II

Is there a link between exploratory behaviour and positive emotions in juvenile female and male pigs (Sus scrofa)?

L. McKenna[1], Ahmad Reza Sharifi[1] and M. Gerken[1]

[1]Department of Animal Sciences, University of Goettingen, Albrecht-Thaer-Weg 3, 37075 Goettingen, Germany

Applied Animal Behaviour Science

Submitted September 2018

Abstract

The aim of the present study was to investigate whether exploratory behaviour and vascular parameters are linked and could be interpreted as coherent indicators of emotional states in growing pigs. Being an essential behaviour for pigs, exploration of their surroundings might have the potential to trigger positive emotions. Our study investigated the effect of eight novel objects on the behaviour and the corresponding vascular responses of 9 female and 9 male growing pigs in postnatal weeks 6-9. The novel objects tested were a rope, a soil heap, a Kong®, an experimental glove, a dog intelligence game, a wallow, a rubber duck and a heap of leaves. Every animal was tested twice per object. During novel object tests, the exploratory behaviour towards the objects and the test environment as well as play and tail wagging behaviour were recorded. Simultaneously, heart rate (HR) and further cardiovascular parameters (RR, SDNN, RMSSD, NN50, LF, HF) were measured using the Polar® System. Results show that "complex" objects (soil, wallow, leaves) were explored more than "rigid" objects (rope, glove, Kong®, IQ game, duck; $p = 0.01$). The novel objects tested had a significant influence on the behavioural traits, HR, RR and LF ($p = 0.001$). Relations between exploratory behaviour and vascular parameters were not always coherent and differed between objects. Across all novel objects, NN50 showed high to medium positive correlations with all tested behavioural parameters. However, the correlations between behavioural traits and the other vascular parameters were mainly not significantly different from zero or only of low magnitude. We conclude from our findings that the suggested link between exploratory behaviour and vascular parameters could be useful for the detection of positive emotions. However, vascular parameters are easily influenced by factors such as locomotion or diurnal rhythm, which might explain the inconsistency of our results.

Keywords

Exploratory behaviour, positive emotions, novel object test, heart rate, heart rate variability, pigs

1. Introduction

It is agreed, that our farm animals are sentient beings, able to experience emotions (Špinka, 2012). Emotions enable survival as animals avoid situations which cause pain or fear and strive towards situations which elicit pleasure and contentment. The presence of positive affective states, and not only the absence of negative affective states, is often described as good welfare (Bradburn, 1969, Seligman and Csikszentmihalyi, 2000), and distinguishes a good quality of life (Morton, 2007).

However, emotional states are only accessible for the individual via introspection. Different concepts have been developed to derive an insight into the animals' affective states using behavioural indicators and physiological changes. After the appraisal of a situation by the individual, physiological changes dictate, as a form of action pattern, how best to behave in a specific situation. The resulting behavioural changes can therefore also be used as a reliable measure to assess emotionality in animals (Paul et al., 2005). On the behavioural level, it could be hypothesized that the active interaction with the environment results in or is caused by the experience of positive emotions. This connection between positive emotions and active engagement with the environment is often called "positivity offset" (Diener and Diener, 1996, Ito and Cacioppo, 1999). Positivity offset prompts individuals to approach novel objects, situations or other individuals (Frederickson and Cohn, 2008) and positive emotions have thus been described as the facilitation of approach behaviour (Davidson, 1993, Cacioppo and Gardner, 1999).

In pigs, exploration is an important behavioural expression. Being omnivorous animals, pigs explore large areas in search for various edible materials which are often sporadically distributed (Studnitz et al., 2007). In domestic pigs kept in a semi-natural environment, Stolba and Wood-Gush (1989) found that 75% of day time was spent with foraging-related activities. If given a choice, pigs preferred to forage or "work" for food, instead of just consuming openly provided food (De Jonge et al., 2008), a phenomenon termed as contrafreeloading. De Jonge et al. (2008) found that pigs favoured entering an environment with straw and hidden chocolate raisins over entering an alternative environment with straw and food at the animal's disposal in a trough. Fraser et al. (1991) showed that growing pigs kept in an environment without substrate redirected their exploratory behavior (rooting and chewing) towards their pen mates, while the control animals which were provided with substrate concentrated their exploratory behaviour towards the given substrate. Such findings underline the suggestion that the expression of exploratory behaviour in pigs represents a behavioural need (Thorpe, 1965).

On the physiological level, imaging techniques (e.g. fMRI, PET scans) enable the detection of changes within the brain accompanying positive emotional states (Burgdorf and Panksepp, 2006; Huettel et al., 2009). Also salivary protein (Morse et al., 1989; Grigoriev et al., 2003) and heart rate variability (HRV) (Fox et al., 1989; McCraty et al., 1995; Rainville et al., 2006; Von Borell et al., 2007) have been investigated to draw inferences about positive emotions. The limbic system, located in sub-neocortical areas of the brain, is largely involved in the generation of emotions.

A main efferent pathway of the limbic system is the autonomic nervous system (ANS) consisting of sympathetic and parasympathetic control mechanisms which correlate with the physiological state of an organism, e.g. stress or homeostasis (Boissy et al., 2007). This physiological state is generally reflected by heart rate (HR) and HRV as the heart activity is both under the control of the sympathetic and parasympathetic branch of the ANS. Due to direct innervation of the heart through sympathetic and parasympathetic fibres, emotions therefore have a large effect on cardiac activity (Saul, 1990). Sympathetic control of heart activity is indicated by an increase of heart beats and decrease of HRV, commonly associated with stress responses and negative emotional states. Parasympathetic control of cardiac activity is marked by lower HR and higher HRV, linked to relaxation and positive emotional states (Rainville et al., 2006; Von Borrell et al., 2007). HRV is therefore considered as potential indicator of the emotional state experienced by an organism (e.g., Appelhans and Luecken, 2006) and was used in several studies in pigs exposed to emotional arousal. In the study of De Jong et al (2000), pigs which had previously won a fight with an unknown conspecific had higher HRV than pigs which had previously lost a fight. Zebunke et al. (2013) found that pigs which solved a cognitive challenge and obtained a food reward showed lower HR and higher HRV compared to pigs fed conventionally.

To improve animal welfare, it is crucial to develop objective methods to assess emotional states of animals (Dawkins, 2008). The aim of the present study was to investigate whether exploratory behaviour and vascular parameters are linked and could be interpreted as coherent indicators of emotional states in growing pigs. To induce exploratory behaviour, different novel objects were introduced. It has been suggested that attractive materials for pigs are characterized by being changeable, destructible as well as edible (Studnitz et al., 2007). Straw is therefore most commonly used in studies evaluating exploratory behaviour in pigs as it fulfills the above mentioned criteria (Kelly et al., 2000; Day et al., 2002; Studnitz et al., 2007). In our study, we decided to test the effect of a variety of objects of a complex or rigid structure on the exploratory behaviour of male and female pigs. During testing, behaviour of the pigs was video-taped and the vascular responses were measured by recording of HR and HRV for later analysis of the relationship between behavioural and physiological responses.

2. Animals, Materials and Methods

2.1 Animals, housing and management

In total, 9 sows and 9 boars (Pietrain x German Landrace) of three different litters were involved, originating from the University's research farm in Relliehausen, Germany. At 28 days of age (average body weight 8.16kg), all animals were weaned and arrived at the research facility of the Department of Animal Sciences, University of Goettingen. The animals had their tails undocked. The pigs were kept in unisexual groups, each group containing animals from all three litters and housed in pens of approximately $18m^2$ each (2 m²/ animal) lined with straw and sawdust. The pens were cleaned daily, fresh straw was provided once weekly and fresh sawdust was provided when necessary. Age appropriate pelleted feed was offered and comprised starter feed with 17.7% CP from week 4 to 6 (Una Hakra, Hamburg, Germany); thereafter the animals received feed mixed at the research farm Relliehausen appropriate for the first fattening phase with 17.6% CP until the end of the experiment. Feed and fresh drinking water were offered *ad libitum* for all animals during the whole experimental period. The animals were provided with daylight and additional artificial light from 07:00 to 19:00 hours. The weight of all animals was monitored weekly (Salter Brecknell, Smethwick, UK, to the nearest 10 g).

2.2 Test room

The test room consisted of three compartments (Figure 1). All compartments were separated by solid wooden barriers. The anticipation compartment (A) was reached through a swing door from the preparation compartment (B). The testing compartment (C) was separated from the anticipation compartment by a guillotine door. Two digital video cameras were installed (Rollei Movieline SD55, Hamburg, Germany). The flooring of all compartments consisted of barren concrete ground. During testing, varying stimuli were distributed on the floor of the testing compartment (C).

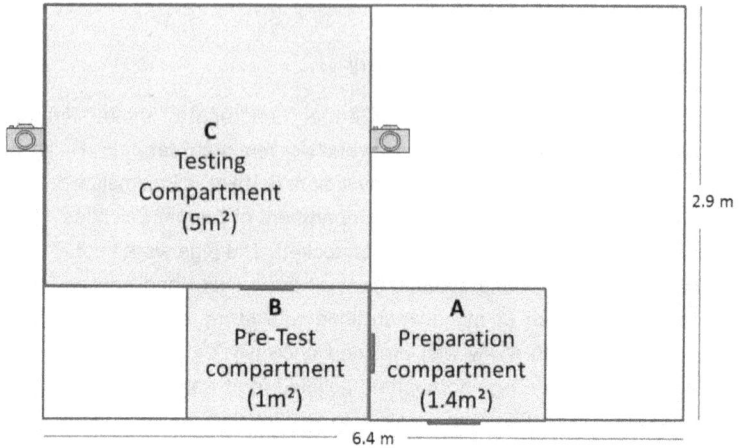

Figure 1: Layout of testroom. Pigs reached the testing compartment (C) via the preparation (A) and the pre-test (B) compartment.

2.3 Testing procedure

After arriving at the test facilities, the animals went through a habituation period of seven days. The experimenter sat in the experimental pens with the animals for approximately one hour daily and casually touched the animals. The animals were also introduced to an elastic belt (daily about 10 Min) around their thorax which was later used for measuring HR and HRV.

Novel object tests were conducted in postnatal weeks 6 to 9. Animals were tested in groups of the same three animals to avoid social isolation stress during testing. Each sex was represented by three groups of the same three animals each, resulting in six groups in total. Eight novel objects were tested (Table 1). Each group was confronted with all novel objects twice in a random order and went through one test per day. In total, 96 tests were conducted involving 3 animals each, resulting in a total of 288 observations. Test duration with the novel objects was 7 minutes. During testing, behaviour was continuously recorded by two video cameras (Rollei Movieline SD55, Hamburg, Germany) with 30 frames/ sec. For measurement of vascular responses the Polar® RS800CX (Electro Oy, Finland) was used. Both integrated electrodes of an elastic belt were coated with electrode gel to enhance conductibility. A sending unit was connected to the elastic belt, transferring the HR data to a receiving unit. Both heart beats and the intervals between heartbeats (RR-Intervals) were recorded. The Polar® system was validated for the use in pigs by Marchant-Forde et al. (2004) and was also used for behaviour tests in young pigs by Zebunke et al. (2011).

Table 1: Novel Objects used in the behaviour tests

Novel Objects	Description
Soil	A heap of potting soil was placed in the centre of the testing arena.
Leaves	A heap of dried leaves was placed in the centre of the testing arena.
Wallow	A wooden frame (163 X 90 cm) was laid out with pondliner and filled with potting soil and water. This wallow was placed in the centre of the testing arena.
Gloves	Three transparent rubber gloves were filled with warm water and laid out on three marked spots in the testing arena.
Ducks	Three yellow rubber ducks (9 cm height) which squeaked when pressed were laid out on three marked spots in the testing arena
Kong®	Three red rubber dog toys "Kong"® (7 cm height) were filled with grapes and laid out on three marked spots in the testing arena.
Intelligence Games (IQ)	Three wooden "intelligence games" (Karlie "Doggy Brain Train") for dogs were filled with grapes and laid out on three marked spots in the test arena. The grapes were hidden in the toy under wooden discs. By pushing around the discs the pigs could reach the grapes.
Ropes	Three hemp ropes (2 cm diameter, 23 cm length) with a knot on each end were laid out on three marked spots in the testing arena.

For the test procedure, all animals were individually marked by a number sprayed on their back. Tests were carried out on 16 test days in the morning (09:00 to 11:00 h) and the afternoon (13:00 to 15:00 h). The sequence of the test groups was randomized across daytime and sex. For testing, the doors of the home pen were opened and all pigs were allowed to proceed into the hallway. From there, test individuals were led into the preparation compartment, while the remaining pigs returned from the hallway to the home pen.

In the preparation compartment the elastic belts of the Polar® system were attached around the thorax of the test animals, directly behind the front legs and the electrodes were placed across the sternum. Time for these procedures was set to one minute for all three individuals. The animals were then led into the pre-test compartment. The experimenter left the test room and the animals remained in the pre-test compartment for one minute. The experimenter re-entered the test room and led the animals further into the testing compartment where the animals found three same items of one of the respective novel objects (Table 1). The experimenter left the test room and re-entered after seven minutes. Pigs were led back through the pre-test into the participation compartment. Elastic belts were taken off and the animals were moved back to their home pen.

2.4 Behavioural Analysis

Video recordings during the novel object tests were individually analyzed for exploratory behaviour of the novel object and the environment by using the time sampling method (Martin and Bateson, 2007). Time intervals were set at 20 seconds resulting in 21 time sampling points per test. Exploratory behaviour was defined as touching, manipulating, chewing or sniffing the object or environment (walls and floor) with the snout. Occurrence of exploratory behavior was then expressed as percentage (%) of time sampling points recorded.

Play behaviour was defined as running, gamboling or pivoting. Tail wagging behaviour was scored when the animals' tail was swinging from side to side. Both traits were sampled continuously and expressed as the total frequency of occurrence. Latency to first touch the novel object was measured in min starting from entering the test arena until the first touch.

2.5 Heart Rate Analysis

HR and HVR recordings during which the connectivity between sending and receiving unit was lost during the testing time, were discarded from further analysis. For further analyses only uninterrupted segments of five min (Sloan et al., 1994; Von Borrell et al., 2007) starting when the animals entered compartment C were included. Thus, of a total of 288 measurements, only 66 (23%) of the heart rate measurements were suitable for analysis. The remaining recordings were corrected using the Kubios Software for HRV analysis (Biosignal and Medical Imaging Group, University of Kuopio, Department of Applied Physics, Kuopio, Finland, for review see Niskanen et al., 2002) applying the software's medium artefact correction level. In addition to HR (beats per minute, bpm), HRV was analyzed in the time domain by using the parameters RR (duration of intervals between consecutive heart beats), SDNN (standard deviation of RR intervals), RMSSD (root mean square of the difference between successive RR intervals reflecting vagal activity only), RMSSD/SDNN (vago-sympathical balance) and NN50 (number of pairs of successive beat-to-beat intervals differing by < 50 ms). HRV was also analyzed in the frequency domain by using LF (low frequency band, 0.04 – 0.33 Hz), HF (high frequency band, 0.33 – 0.83 Hz), and LF/HF (sympatho-vagal balance; Von Borell et al., 2007).

2.6 Statistical Analysis

Statistical analyses were performed using SAS (SAS 9.3, SAS Institute Inc., 2010). Behavioural traits measured were not normally distributed. Data on exploration of environment and object were expressed as percentage (%) of time sampling points recorded and binomially distributed. For analyses, a generalized linear model using the GLIMMIX procedure was performed using the following model:

$$\mathrm{logit}(p_{ijk}/1\text{-}p_{ijk}) = p_{ijk} = \varphi + \alpha_i + \beta_j + \alpha\beta_{ij} + \lambda_k \quad (I)$$

where p_{ijk} is the probability of the animals to explore the environment or the object, φ is the overall mean effect; α_i is the fixed effect of gender (male, female); β_j is the fixed effect of object (rope, soil, kong, glove, IQ, wallow, duck, leaves); $(\alpha\beta)_{ij}$ is the interaction between the fixed effects and λ_k random effect of animal due to repeated measurements.

Least square means were estimated on the logit scale and then back-transformed using the inverse link function $\pi = \exp(x\beta)/[1 + \exp(x\beta)]$ to the original scale (probability) applying the LSMEANS statement.

Statistical analysis of play and tail wagging data were carried out using the following generalized linear model underlying a poisson distribution:

51

$$\eta_{ijk} = \log(y_{ijk}) = \mu + \alpha_i + \beta_j + \alpha\beta_{ij} + \lambda_k \quad \text{(II)}$$

μ is the overall mean effect; α_i is the fixed effect of gender; β_j is the fixed effect of object; $(\alpha\beta)_{ij}$ is the interaction between the fixed effects and λ_k is the random effect of animal due to repeated records. Least square means were estimated on the log scale and then back-transformed using the inverse link function $\exp(\eta_{ijk})$ to the original scale (applying the LSMEANS statement).

During the novel tests, 19 animals took more than 30 seconds to first touch the object These data points were excluded as outliers from further analysis of latency time. Log transformation was carried out and statistical analysis was carried out using the following generalized linear model underlying a normal distribution:

$$y_{ijkl} = \mu + \alpha_i + \beta_j + \alpha\beta_{ij} + \lambda_k + e_{ijkl} \quad \text{(III)}$$

μ is the overall mean effect; α_i is the fixed effect of gender; β_j is the fixed effect of object; $(\alpha \times \beta)_{ij}$ is the interaction between the fixed effects, λ_k is the random effect of animal due to repeated recordings and e is the random error.

The same model (III) as described for latency time was used for the analysis of the untransformed vascular parameters HR, RR, SDNN, RMSSD, RMSSD/SDNN, pNN50, LF, HF and LF/HF underlying a normal distribution.

Significant differences between least square means were tested using a tukey multiple t-test. Standard errors of least square means were calculated as described by Littell et al. (1999). Statistical significance was set at $P < 0.05$. For a better understanding, results are shown as retransformed data.

For analysis of the relation between exploratory behaviour and vascular parameters Kendall rank correlation coefficients (Kendall's tau coefficient, τ) were calculated using SAS (SAS 9.3, SAS Institute Inc., 2010).

3. Results

For the behavioural traits recorded, no significant differences between sexes were found (Table 2). Females, however, tended to show more exploratory behaviour than males ($p = 0.067$). The various objects provoked considerably different reactions for all traits ($p < 0.0001$) with the exception of latency to touch the object ($p = 0.134$). The pigs explored the environment significantly less when the objects soil, wallow and leaves were tested. On the other hand, these objects were also the ones which were explored most of the time. The food-baited objects (Kong® and IQ) resulted in medium object exploration, while the objects duck, rope and glove received the least exploration. Significant interactions between gender and object were found for exploring object, play and tail wagging behaviour (Figure 2). The males explored the wooden (IQ) and rubber (Kong®) dog toys more than the females (males: 46%, 52.2% vs females: 36.2%, 34.1%, respectively), whereas the females explored the wallow to a larger extent (males: 60.6% vs females: 74.2%).

Table 2: The effects of gender, object and their interaction on different behaviour and cardio-vascular traits.

Traits	Significance of effects, P-value		
	Gender	Object	Gender*Object
Exploring Environment	0.067	<0.001	0.351
Exploring Object	0.225	<0.001	0.006
Play	0.990	<0.001	0.006
Tail Wagging	0.161	<0.001	<0.001
Latency Object	0.174	0.134	0.207
HR	0.130	<0.001	0.039
RR	0.212	<0.001	0.017
RMSSD	0.022	0.104	0.557
SDNN	0.256	0.326	0.493
RMSSD/SDNN	0.133	0.063	0.264
NN50	0.074	0.161	0.150
LF	0.868	0.018	0.916
HF	0.094	0.331	0.320
LF/HF	0.160	0.563	0.735

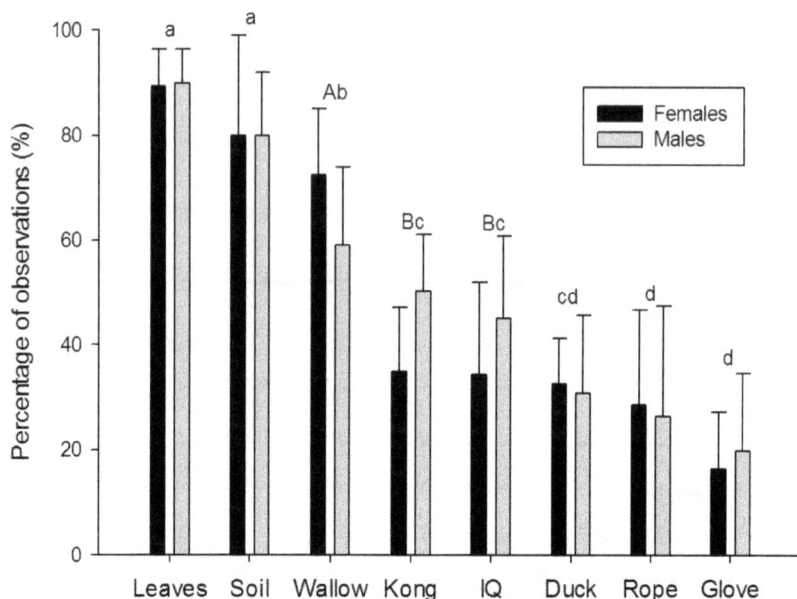

Figure 2: Percentage of observations (%) of object related exploratory behavior (means ± SD) in male (n=9) and female (n=9) pigs across different objects shown during novel object tests. For explanation of novel objects see Table 1.
[ab] significant differences between objects, P<0.05
[AB] significant differences between sexes within the same object, P<0.05.

Rope, duck and Kong® elicited most play behaviour, while least play was shown during tests with wallow, IQ and leaves (Table 3). Females exhibited most play behaviour during tests with the novel object rope (1.83 times), males displayed most play behaviour in presence of the Kong® (1.61 times). The highest occurrence of tail wagging was observed when the duck and wallow were tested. Females showed most tail wagging when the rope was presented (1.83 times) and males were wagging most frequently when the Kong® was tested (1.61 times). All pigs touched the objects during the different testing situations, but latency time was not effected by gender, object or the interaction thereof.

Table 3: LS-means (±SEM) for exploring environment and object (% of observations), play and tail wagging behaviour (frequency), and latency to touch object (min), by gender and object.

Effect	Exploring environment (% of observations)[1]	Exploring object (% of observations)[1]	Play (frequency)	Tail wagging (frequency)	Latency to touch object (min)
			Trait		
Gender (G)					
Female	19.1 ± 1.8^a	48.9 ± 1.7^a	0.66 ± 0.16^a	1.82 ± 0.50^a	0.12 ± 0.02^a
Male	14.2 ± 1.8^a	51.9 ± 1.7^a	0.72 ± 0.17^a	0.78 ± 0.20^a	0.16 ± 0.05^a
Object (O)					
Rope	31.2 ± 3.0^a	24.0 ± 2.7^{fe}	1.67 ± 0.62^a	0.89 ± 0.63^{abc}	0.09 ± 0.01^a
Soil	3.8 ± 1.3^b	77.6 ± 2.8^b	0.75 ± 0.34^{abd}	0.54 ± 0.26^a	0.08 ± 0.01^a
Kong	28.6 ± 2.6^a	42.9 ± 2.7^c	1.00 ± 0.44^{abd}	0.53 ± 0.50^a	0.08 ± 0.01^a
Glove	42.9 ± 2.7^c	19.3 ± 2.1^f	0.33 ± 0.14^{bc}	1.69 ± 0.69^{bc}	0.05 ± 0.01^a
IQ	26.8 ± 2.6^a	41.0 ± 2.7^{cd}	0.18 ± 0.08^{bc}	0.79 ± 0.44^{ab}	0.08 ± 0.01^a
Wallow	14.9 ± 2.0^c	67.8 ± 2.5^b	0.16 ± 0.09^c	2.00 ± 0.52^c	0.06 ± 0.01^a
Duck	25.0 ± 2.5^a	31.4 ± 2.6^{de}	1.21 ± 0.32^{ad}	2.61 ± 1.57^c	0.06 ± 0.01^a
Leaves	1.9 ± 1.1^b	89.6 ± 2.3^a	0.28 ± 0.14^{bcd}	1.00 ± 0.31^{abc}	0.07 ± 0.01^a

a-d Means within effects with different superscripts differ significantly (P < 0.05).
[1] Back-transformed least squares means.

The effects of gender, object and the interactions thereof on vascular parameters are summarized in Tables 4 and 5. HR was not influenced by gender (p = 0.13) but by the novel object (p = 0.01). Highest HR was observed when the novel object duck was tested and leaves resulted in the lowest HR. The sex x object interaction was significant (p = 0.039). While the highest HRs were shown in both genders when the duck was tested (females: 170.9bpm, males: 168.78 bpm, respectively), lowest HRs were measured with glove for females (148.01 bpm) but with leaves for males (143.95bpm).

Table 4: Time domain analysis of HRV. LS-means (±SEM) for HR (bpm), RR (ms), RMSSD (root mean square of successive beat-to-beat differences), SDNN (standard deviation of beat-to-beat intervals), RMSSD/SDNN (Vago-Sympathical Balance) and NN50 (Normal to Normal <50ms) by gender and object.

Effect	Trait					
	HR[1]	RR[2]	RMSSD[3]	SDNN[4]	RMSSD/SDNN[5]	NN50[6]
Gender (G)						
Female	158.43 ± 1.85[a]	379.92 ± 5.42[a]	6.05 ± 0.47[a]	16.52 ± 1.47[a]	0.37 ± 0.03[a]	0.87 ± 1.15[a]
Male	162.20 ± 1.57[a]	370.25 ± 5.30[a]	7.52 ± 0.39[b]	18.72 ± 1.19[a]	0.43 ± 0.03[a]	3.62 ± 0.93[a]
Object (O)						
Rope	160.17 ± 2.63[ab c]	371.22 ± 7.28[ab]	7.71 ± 0.69[a]	15.89 ± 2.05[a]	0.52 ± 0.44[a]	1.65 ± 1.61[a]
Soil	168.23 ± 4.81[ab]	360.30 ± 11.95[ab c]	4.76 ± 1.26[a]	17.01 ± 3.75[a]	0.33 ± 0.08[a]	0.27 ± 2.94[a]
Kong	153.42 ± 2.74[ac]	391.42 ± 7.86[ac]	7.10 ± 0.72[a]	12.89 ± 2.92[a]	0.50 ± 0.04[a]	1.18 ± 2.29[a]
Glove	153.65 ± 3.75[ac]	387.90 ± 9.31[a]	5.87 ± 0.98[a]	17.75 ± 2.92[a]	0.35 ± 0.06[a]	0.34 ± 2.29[a]
IQ	163.88 ± 3.73[ab c]	363.62 ± 9.23[ab]	6.17 ± 0.98[a]	17.27 ± 2.91[a]	0.36 ± 0.06[a]	6.66 ± 2.28[a]
Wallow	163.51 ± 2.47[ab]	369.56 ± 6.23[ab]	7.54 ± 0.65[a]	21.10 ± 1.92[a]	0.36 ± 0.04[a]	1.46 ± 1.51[a]
Duck	169.84 ± 2.97[b]	350.71 ± 7.43[b]	6.56 ± 0.78[a]	18.56 ± 2.31[a]	0.37 ± 0.05[a]	0.72 ± 1.81[a]
Leaves	149.80 ± 2.33[c]	405.92 ± 5.90[c]	8.58 ± 0.61[a]	20.49 ± 1.82[a]	0.44 ± 0.04[a]	5.67 ± 1.42[a]

[a-d] Means within effects with different superscripts differ significantly ($P < 0.05$).

[1] Heartbeats per minute.

[2] Beat-to-beat interval in ms.

[3] Root mean square of successive differences between successive beat-to-beat intervals.

[4] Standard deviation of beat-to-beat intervals.

[5] Vago-sympathical balance

[6] Number of pairs of successive beat-to-beat intervals differing by more than 50ms

Table 5: Frequency domain analysis of HRV. LSQmeans (±SEM) for LF (low frequency band), HF (high frequency band) and LF/HF (sympatho-vagal balance) by gender and object.

Effect	Trait		
	LF[1]	HF[2]	LF/HF[3]
Gender (G)			
Female	98.11 ± 27.39[a]	7.78 ± 3.96[a]	16.63 ± 3.33[a]
Male	104.03 ± 22.13[a]	16.59 ± 3.19[a]	9.81 ± 3.36[a]
Object (O)			
Rope	68.81 ± 38.11[a]	11.34 ± 5.51[a]	12.67 ± 3.08[a]
Soil	48.84 ± 69.63[ab]	3.88 ± 10.07[a]	9.78 ± 5.08[a]
Kong	51.99 ± 54.28[ab]	10.72 ± 7.85[a]	13.00 ± 3.95[a]
Glove	86.47 ± 54.28[ab]	6.27 ± 7.85[a]	15.67 ± 4.06[a]
IQ	82.03 ± 54.12[ab]	18.80 ± 7.83[a]	12.62 ± 3.96[a]
Wallow	258.43 ± 35.69[b]	16.69 ± 5.16[a]	16.84 ± 2.98[a]
Duck	87.66 ± 42.99[ab]	6.09 ± 6.22[a]	14.22 ± 3.37[a]
Leaves	124.30 ± 33.67[ab]	23.71 ± 4.86[a]	10.96 ± 2.87[a]

[ab] Means within effects with different superscripts differ significantly (P < 0.05).

[1] Low frequency band (0.04 – 0.33 Hz) in ms^2

[2] High frequency band (0.33 – 0.83 Hz) in ms^2

[3] Sympathovagal balance

Similarly to HR, the effects of object and the interaction between gender and object were significant for RR measurements. For RMSSD, gender had an influencing effect (p = 0.022) with females showing a lower value than males. The effects of object or the interactions were not significant. The other parameters in the time domain analysis (SDNN, RMSSD/SDNN, NN50) were not influenced by the tested effects (Tab. 2 and 4). However, RMSSD and RMSSD/SDNN tended to be influenced by objetct (p =0.104 and 0.063, respectively). In the frequency domain analysis (Tab. 2 and 5), LF was significantly influenced by object with significant (p = 0.018) differences between wallow and rope (258.43 vs rope: 68.81ms²). The other parameters of the frequency domain analysis (HF and LF/HF) were not affected by the tested effects.

The course of HR during the test is illustrated for the examples leaves and duck, the novel objects resulting in the most diverse reactions (Fig. 3). While object related exploratory behaviour remained at a high level throuout the test for the leaves (average 89.6%), exploration of the ducks started at a high level and quickly declined within the first minute (average 31.4%). When the animals were confronted with the leaves, vascular reactions resulted in a low HR, but vice versa when the duck was presented as novel object (Fig. 3a and b).

Among behavioural traits a close negative correlation was found between exploring environment and exploring object (Table 6), while the other correlations between behavioural traits were close to zero or of only low magnitude. The correlations between behavioural traits and the vascular parameters of the time domain (HR, RR, RMSSD, SDNN) were mainly not significantly different from zero or only of low magnitude (Table 6). Similarly, correlations between cardio-vascular traits of the frequency domain and behavioural traits were not significantly different from zero. In contrast, NN50 showed high to medium positive correlations with all tested behavioural parameters.

Among vascular traits, SDNN was highly positive correlated with HR, RR and LF. NN50 showed high correlations with the behavioural parameters explore environment, explore object and play (Table 6).

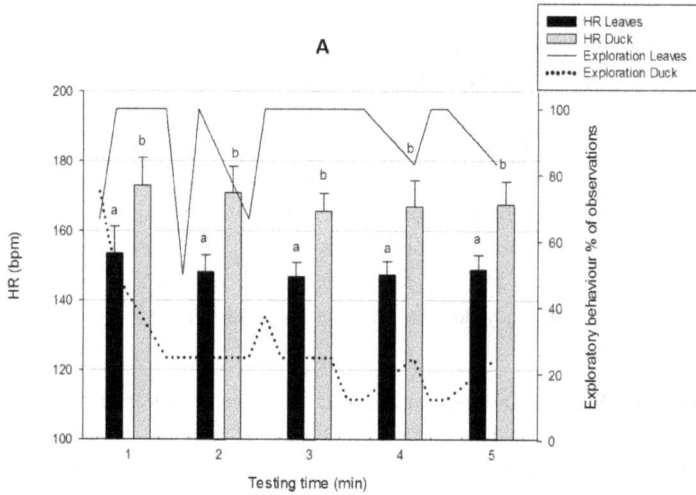

Figure 3: Development of heart rate (HR, bpm) and object exploratory behaviour during the course of the analysed testing time (5 min) for the novel objects leaves and duck, means across 18 animals of both sexes ± SD. Different small letters indicate significant differences in heart parameters between objects, p≤0.05.

Table 6: Kendall's tau coefficients (τ) between behavioural and cardio-vascular traits (N = 66 measurements).

Trait	Explore Object	Play	Tail Wagging	Latency touch obj.	HR	RR	RMSSD	SDNN	NN50	LF	HF	LF/HF
Explore Environm.	-0.543***	0.107*	-0.013	0.031	-0.046	0.112	0.049	0.011	0.826***	-0.08	-0.021	-0.024
Explore Object		-0.107†	-0.017	-0.001	-0.019	-0.024	-0.007	0.137	0.810***	-0.09	-0.019	0.002
Play			-0.013	0.157**	0.028	0.049	-0.078	0.197**	0.614***	-0.06	-0.011	-0.043
Tail Wagging				-0.100	0.084	-0.129	0.171†	-0.005	0.261**	-0.08	-0.030	-0.041
Latency touch obj.					-0.102	0.077	0.006	0.106	0.296**	-0.09	-0.052	-0.045
HR						-0.690***	-0.113	0.696***	-0.409***	0.601***	0.049	-0.012
RR							0.024	0.845***	-0.353***	0.659***	0.028	-0.031
RMSSD								0.048	0.261**	-0.057	0.089	0.091
SDNN									-0.232**	0.728***	0.088	0.016
NN50										-0.281**	-0.001	0.015
LF											0.168*	-0.011
HF												0.897***

† p < 0.1; * p < 0.05 ; ** p < 0.01; *** p < 0.00

4. Discussion

We induced exploratory behaviour in juvenile pigs by novel object tests and investigated the possible link between their exploratory reactions and vascular parameters. Thus, the aim was to evaluate if traits measured at the behavioural and physiological level could be interpreted as coherent indicators of emotional states in growing female and male pigs.

For the behavioural traits recorded, we found no differencves between sexes. However, the tendency of increased environmental exploration in females is in line with the results of Bolhuis et al (2005) who found more exploration behaviour in gilts than in barrows (p < 0.05). As the animals in that study (Bolhuis et al., 2005) were tested at a later age (20 weeks of age), more distinctive sex differences might emerge at a later developmental stage. Similarly, the present sex x object interactions indicate possible differences between sexes with regard to attractivity of specific object properties.

Pigs of both sexes preferred objects of higher complexity and plasticity (leaves, wallow and soil) over rigid objects. Similarly, Scott et al. (2007), found that pigs primarily chose organic material with higher plasticity, which they explored around 20% of the time, compared to an artificial toy which was only explored around 1% of the time, independent whether the animals were housed on straw-bedded or slatted floor pens. This strong preference for materials of higher complexity (mainly of organic origin) and changeability could be attributed to the broader range of olfactory and tactile stimuli which may be inherently associated with natural foraging behaviour. Jensen and Pedersen (2007) found that among organic rooting materials, pigs prefer more complex and compound materials (e.g. maize silage with straw) over more homogenous materials (e.g. chopped straw). In our study, pigs might have explored leaves more than soil or wallow for a similar reason. The home environment of our animals was enriched with sawdust and straw. It is noteworthy that even though pigs had permanent access to complex rooting material, novel materials of high changeability were favoured in the novel object tests. Interestingly, pig also increased their exploration of the environment when less attractive objects were presented as shown bei the negative correlation between both traits

The chosen stimuli in the novel objetc tests elicited different arousals of the animals. As all animals touched the objects, an underlying positivity offset (Diener and Diener, 1996, Ito and Cacioppo, 1999; Frederickson and Cohn, 2008) can be assumed. At the behavioural level, positive emotional states can be expressed through play (Špinka et al., 2001; Paul et al., 2005; Boissy et al., 2007; Held and Špinka, 2011) or tail postures (Kiley-Worthington, 1976; Reefmann et al., 2009). Reimert et al. (2013)

found that play behaviour and TW occurred significantly more often when the test pigs (aged 12 weeks) were given a positive treatment (access to an area filled with peat, straw and chocolate raisins). This finding is in line with early literature observing piglet play in semi-natural surroundings and coinciding play and TW (Newberry et al., 1988). In the present study, we found a low positive correlation (τ = 0.107) between exploratory and play behaviour which might reflect a general state of increased positivity offset. Accordingly, play behaviour could indicate positive affective states during exploration of the test situation. On the other hand, pigs exhibiting more play also had an slightly increased latency to touch the object (τ = 0.157) .

In our study, the novel objects elicited significant differences in play and TW behaviour. The results, however, are difficult to interpret. Male animals played and wagged their tails most when the Kong® was tested (1.61 times each). As this object was food-baited, the presence of the food might have acted as trigger. Searching for food is an essential behaviour in pigs, so that they are even choosing to root for hidden food, instead of eating equal food provided from a trough (De Jonge, 2008). However, the object IQ was equally baited, but resulted in low overall play behaviour. Reduced play behaviour was also observed when wallow and leaves were tested. We suggest, that the intensified exploratoration of the attractive objects of high complexity superimposed play behaviour. As shown for the objects leaves and duck, exploration activitiy is shifted according to the acctractivity of the object, e.g. to the environment, when the object is less attractive.

We hypothesized that the objects for which the pigs showed highest exploration, also elicited higher HRV and consequently lower HR (Rainville et al., 2006; Von Borrell et al., 2007). Zebunke et al. (2013) demonstrated that pigs reacted with high HRV and low HR to situations of positive valence (solving a cognitive challenge and individually gaining a food reward) compared to a situation of negative valence (conventional feeding accompanied by agonistic interactions with conspecifics) which resulted in increased HR and decreased HRV. Similar results were observed in sheep (Reefmann et al., 2009) when the reactions were evaluated in situations of positive, neutral and negative valence, the RMSSD value of the sheep also corresponded to these findings as it was highest in the situation of positive valence (Reefmann et al., 2009).

In our study, we found low HR and high RR for the leaves among the objects which were explored at a high frequency, and the opposite for the duck representing the less intensively explored objects. These findings may suggest an underlying positive emotional state when the pigs explored the leaves and a less positive valence for the animals confronted with the duck. There was also a tendency for highest RMSSD values during exploration of the leaves. The HRV parameter LF (low frequency band of the power spectrum) was found to be higher when the animals explored the wallow when compared to the rope. While the HF component is exclusively associated with the parasympathetic branch of the ANS, the LF component encompasses both parasympathetic and sympathetic elements. Therefore the physiological meaning of the LF component is much debated, as it can easily be influenced by other physiological measures such as thermoregulation and myogenic activity of vessels (Ponikowski et al., 1996; Van Borell et al., 2007). We have found a highly significant negative correlation between HR and RR. This finding was expected from the previous results on the interdependence between these traits due to the sympathetic and parasympathetic control of cardivascular reactions (Rainville et al., 2006, von Borell et al., 2007).

Interestingly, even though the tested effects (object and gender) did not consistently influence the HRV parameters, we have found high positive correlations between NN50 and behavioural parameters indicative for positive emotion (exploration, play, tail wagging). As NN50 measures beat-to-beat intervals exceeding 50ms in length, this parameter is indicative for a high HRV, therefore an activated parasympathic system.

A correspondingly elevated RMSSD value, which would also have confirmed a positive valence when the animals were showing explorative, play and tail wagging behaviour however, was not found. A rise in RMSSD is commonly associated with activation of the parasympathic nervous system (Boissy et al., 2007). Also Zebunke et al. (2013) found that pigs reacted with an elevated RMSSD value to being called for feeding and explained their finding with a positively valenced anticipation of the food.

The lack of coherent correlations between behaviourial and vascular parameters could explained by the low and unbalenced number of measurements of heart parameters available, thus limiting further conclusions. In addition, the interpretation of HR and HRV as indicators of underlying emotional states is not always straightforward and results can be influenced by various factors (Paul et al., 2005). As Pagani et al. (1995) showed, HR and HRV are easily influenced by locomotor activity. In humans, a shift towards sympathetic balance is already taking place when moving from a lying to a standing position (Pagani et al., 1995). Novel object tests in horses showed that physical activity largely accounted for increase in HR (Visser et al., 2002). Also anticipation of positive or negative upcoming events can cause increases in HR (Baldock and Sibly, 1990). In addition, physiological indicators for emotions may be also superimposed by diurnal variations (Schrader and Ladewig, 1999, von Borell and Ladewig, 1992). Kuwahara et al. (1999) showed that HR and HRV in pigs shifts from a vagal to sympathetic dominance over the course of late morning to early afternoon. This shift might have also influenced the HR and HRV of our pigs during the novel object tests as our animals were tested both in the morning and the afternoon in randomized order.

We conclude from our findings that the pigs predominately explored the complex objects and that rigid toys are less suitable than changeable materials as occupational material for pigs. We were able to show some relations between exploratory behaviour and heart rate parameters. However, not for all tested objects a coherent explanation of resulting cardivascular parameters could be found. These observations warrant further research, especially taking the influence of locomotor activity and diurnal patterns on HR parameters into account. In our study, only a limited number of HR data was suitable for analysis. This might explain why a conclusive coherence between the level of exploratory behaviour and the development of HR and HRV was not found. Further research is required to confirm the suggested link between exploratory behaviour and vascular parameters which can be useful for the detection of positive emotional states in pigs.

References

Appelhans, B. M., Luecken, L. J. 2006. Heart rate variability as an index of regulated emotional responding. Rev. Gen. Psychol. 10, 229.

Baldock, N. M., & Sibly, R. M. 1990. Effects of handling and transportation on the heart rate and behaviour of sheep. Appl. Anim. Behav. Sci. 28, 15-39.

Boissy, A., Manteuffel, G., Jensen, M.B., Oppermann, M. R., Spruijt, B., Keeling, L.J., Winckler, C., Forkman, B., Dimitrov, I., Langbein, J., Bakken, M., Veissier, I., Aubert, A. 2007. Assessment of positive emotions in animals to improve their welfare. Physiol. Behav. 92, 375-397.

Bolhuis, J.E., Schouten, W.G.P., Schrama, J.W., Wiegant, V.M. 2005. Behavioural development of pigs with different coping characteristics in barren and substrate enriched housing conditions. Appl. Anim. Behav. Sci. 93, 212-228.

Bradburn N.M. 1969. The structure of psychological well-being. Chicago: Aldline Publishing Company, USA.

Burgdorf, J., Panksepp, J. 2006. The neurobiology of positive emotions. Neurosci. Biobehav. Rev. 30, 173-187.

Cacioppo, J. T., Gardner, W. L. 1999. Emotion. Annu. Rev. Psychol. 50, 191-214.

Davidson, R.J. 1993. The neuropsychology of emotion and affective style, in: Lewis, M., Haviland, J.M. (Eds.), Handbook of Emotions. New York: Guilford Press, 143-154.

Dawkins, M. S. 2008. The science of animal suffering. Ethology 114, 937-945.

Day, J.E.L., Kyriazakis, I., Lawrence, A.B. 1995. The effect of food deprivation on the expression of foraging and exploratory behavior in the growing pig. Appl. Anim. Behav. Sci. 42, 193-206.

Diener, E., Diener, C. 1996. Most people are happy. Psychol. Sci. 7, 181–185.

Diener, E., Seligman, M.E.P. 2004. Beyond money: Toward an economy of well-being. Psychol. Sci. Public Interest 5, 1–31.

Fox, N.A. 1989. Psychophysiological correlates of emotional reactivity during the first year of life. Develop. Psychol. 25, 364-372.

Jensen, M. B., Studnitz, M., Halekoh, U., Pedersen, L. J., Jørgensen, E. 2008. Pigs' preferences for rooting materials measured in a three-choice maze-test. Appl. Anim. Behav. Sci.112, 270-283.

Jong, I. D., Sgoifo, A., Lambooij, E., Korte, S. M., Blokhuis, H. J., Koolhaas, J. M. 2000. Effects of social stress on heart rate and heart rate variability in growing pigs. Can. J. Anim. Sci. 80, 273-280.

de Jonge, F. H., Tilly, S. L., Baars, A. M., Spruijt, B. M. 2008. On the rewarding nature of appetitive feeding behaviour in pigs (Sus scrofa): do domesticated pigs contrafreeload?. Appl. Anim. Behav. Sci. 114, 359-372.

Fraser, D., Phillips, P. A., Thompson, B. K., Tennessen, T. 1991. Effect of straw on the behaviour of growing pigs. Appl. Anim. Behav. Sci. 30, 307-318.

Frederickson, B.L., Cohn, M.A. 2008. Positive Emotions, in: Lewis, M., Haviland-Jones, J.M., Feldmann-Barrett, L. (Eds.), Handbook of Emotions (3rd Ed.), New York: Guilford Press, pp. 777-797.

Grigoriev, I. V., Nikolaeva, L. V., & Artamonov, I. D. 2003. Protein content of human saliva in various psycho-emotional states. Biochemistry (Moscow), 68, 405-406.

Hill, J.D., McGlone, J.J., Fullwood, S.D., Miller, M.F. 1998. Environmental enrichment influences on pig behavior, performance and meat quality. Appl. Anim. Behav. Sci. 57, 51–68.

Huettel, S.A., Song, A.W., McCarthy, G. 2009. Functional Magnetic Resonance Imaging (2.ed.), Massachusetts: Sinauer

Inglis,F.M. J.C. Day, H.C. Fibiger 1994. Enhanced acetylcholine release in hippocampus and cortex during the anticipation and consumption of a palatable meal J. Neurosci. 62, 1049–1056.

Ito, T.A., Cacioppo, J.T. 1999. The psychophysiology of utility appraisals, in: Kahneman, D., Diener, E., Schwartz, N. (Eds.), Well-being: Foundations of hedonic psychology. New York: Russell Sage Foundation, 470–488.

Jensen, M. B., Pedersen, L. J. 2007. The value assigned to six different rooting materials by growing pigs. Appl. Anim. Behav. Sci. 108, 31-44.

Jensen, M. B., Studnitz, M., Halekoh, U., Pedersen, L. J., Jørgensen, E. 2008. Pigs' preferences for rooting materials measured in a three-choice maze-test. Appl. Anim. Behav. Sci. 112, 270-283.

Kahneman, D., Kreuger, A. B., Schkade, D. A. 2004. A survey method for characterizing daily life experience: The day reconstruction method. Science 306, 1776–1780.

Kelly, H.R.C., Bruce, J.M., English, P.R., Fowler,V.R., Edwards, S.A. 2000. Behaviour of 3 week weaned pigs in Straw- Flow, deep straw and flatdeck housing systems. Appl. Anim. Behav. Sci. 68, 269–280.

Kuwahara, M., Suzuki, A., Tsutsumi, H., Tanigawa, M., Tsubone, H., Sugano, S. 1999. Power spectral analysis of heart rate variability for assessment of diurnal variation of autonomic nervous activity in miniature swine. Comp. Med. 49, 202-208.

Martin, P., Bateson, P. 1986. Measuring Behaviour – An introductory guide. Cambridge University Press, UK

McCraty, R., Atkinson, M., Rein, G., Watkins, A. D. 1996. Music enhances the effect of positive emotional states on Salivary IgA. Stress Med. 12, 167–175

Morrison, R.S., Johnston, L.J., Hilbrands, A.M. 2007. The behaviour, welfare, growth performance and meat quality of pigs housed in a deep-litter, large group housing system compared to a conventional confinement system. Appl. Anim. Behav. Sci. 103, 12-24.

Morton D.B. 2007. A hypothetical strategy for the objective evaluation of animal well-being and quality of life using a dog model. Anim. Welf. 16, 75-81.

Morse, D.R., Schacterle, G.R., Furst, M.L., Bose, K. 1981. Stress, relaxation and saliva: A pilot study involving endodontic patients. Oral Surgery, Oral Medicine, Oral Pathology 52, 308-313.

Pagani, M., Lucini, D., Rimoldi, O., Furlan, R., Piazza, S., Biancardi, L. 1995. Effects of physical and mental exercise on heart rate variability. Heart rate variability, 245-266.

Panksepp, J. 2005. Affective consciousness: Core emotional feelings in animals and humans. Conscious.Cogn. 14, 30-80.

Paul, E.S., Harding, E.J., Mendl, M. 2005. Measuring emotional processes in animals: the utility of a cognitive approach. Neurosci. Biobehav. Rev. 29. 469-491.

Ponikowski, P., Chua, T.P., Amadi, A.A., Piepoli, M., Harrington, D., Volterrani, M. 1996. Detection and significance of a discrete very low frequency rhythm in RR interval variability in chronic congestive heart failure. Am. J. Cardiol. 77, 1320.

Rainville P., Bechara A., Naqvi N., Damasio A.R., 2006. Basic emotions are associated with distinct patterns of cardiorespiratory activity. Internat. J. Psychophysiol. 61, 5–18.

Reefmann, N., Wechsler, B., Gygax, L. 2009. Behavioural and physiological assessment of positive and negative emotion in sheep. Anim. Behav. 78, 651-659.

Reimert, I., Bolhuis, J. E., Kemp, B., & Rodenburg, T. B. 2013. Indicators of positive and negative emotions and emotional contagion in pigs. Physiol. Behav. 109, 42-50.

Saul, J.P. 1990. Beat-to-beat variation of heart rate reflects modulation of cardiac autonomic outflow. J. Physiol. 5, 32–7.

Scott, K., Taylor, L., Gill, B.P., Edwards, S. 2007. Influence of different types of environmental enrichment on the behaviour of finishing pigs in two different housing systems: 2. Ratio of pigs to enrichment. Appl. Anim. Behav. Sci. 105, 51-58.

Seligman M.E.P., Csikszentmihalyi M. 2000. Positive psychology: an introduction. Am. Psychol. 55, 5-14.

Sloan, R. P., Shapiro, P. A., Bagiella, E., Myers, M. M., Bigger, J. T., Steinman, R. C., Gorman, J. M. 1994. Brief interval heart period variability by different methods of analysis correlates highly with 24 h analyses in normals. Biol. Psychol. 38, 133-142.

Špinka, M. 2009. Behaviour of pigs, in: Jensen, P. (Ed.) The Ethology of Domestic Animals: An Introductory Text. CAB International, Wallingford, UK, 177-191.

Špinka, M. 2012. Social dimensions of emotions and its implication for animal welfare. Appl. Anim. Behav. Sci. 138, 170-181.

Stolba, A., & Wood-Gush, D. G. M. 1989. The behaviour of pigs in a semi-natural environment. Anim. Prod. Sci. 48, 419-425.

Studnitz, M., Jensen, M. B., Pedersen, L. J. 2007. Why do pigs root and in what will they root?: A review on the exploratory behaviour of pigs in relation to environmental enrichment. Applied Anim. Behav. Sci. 107, 183-197.

Thorpe, W.H. 1965. The assessment of pain and distress in animals. Appendix III in: Report of the Technological Committee to Enquire into the Welfare of Animals Kept Under Intensive Husbandry Condition, F.W.R. Brambell (chairman). HMSO, London.

Von Borell, E., Langbein, J., Després, G., Hansen, S., Leterrier, C., Marchant-Forde, J., Marchant-Forde, R., Minero, M., Mohr, E., Prunier, A., Valance, D., Veissier, I. 2007. Heart rate variability as a measure of autonomic regulation of cardiac activity for assessing stress and welfare in farm animals – A review. Physiol. Behav. 92, 293-316.

Zebunke, M., Puppe, B., Langbein, J. 2013. Effects of cognitive enrichment on behavioural and physiological reactions of pigs. Physiol. Behav. 118, 70-7.

CHAPTER III

Environmental enrichment and emotional states in growing pigs (*Sus scrofa*)

L. McKenna[1] and M. Gerken[1]

[1]Department of Animal Sciences, University of Goettingen, Albrecht-Thaer-Weg 3, 37075 Goettingen, Germany

Manuscript prepared for submission in:

Animal

Abstract

The behaviour of pigs housed in stimulus and substrate enriched environments differs from animals housed in barren conditions because species specific behaviour can be displayed more frequently. The aim of this study was to investigate whether variations in housing enrichment would also affect pigs' behavioural reaction in a novel object test situation. 36 female growing pigs were housed in three different housing environments (n = 12 animals in each environment). The three housing environments were: 1. non-enriched (NE), 2. enriched with substrate and straw (E) and 3. enriched with substrate, straw and weekly changing extra stimuli i.e. grass cut, twigs, vegetables, hay; SE). Animals of each environment were divided into a) experimental animals, which were confronted with novel objects in the test arena and b) controls, which were confronted with an open-field situation (empty test arena). The novel objects used for the experimental animals were soil, wallow, Kongs® and rubber ducks. In this experiment, the effects of treatment, environment and object on the play and tail wagging behaviour and the corresponding cardiovascular parameters of the animals were measured. Play and tail wagging behaviour were additionally summarized as positive emotion score (PES). Results show that the cardiovascular parameters were not significantly influenced by environment or treatment. The experimental animals showed more tail wagging and play behaviour as well as a higher PES when the wallow was tested (1.0 ± 4.8, 6.0 ± 4.8 and 3.5 ± 1.0, respectively) compared to the rubber ducks (0.0 ± 1.4, 0.0 ± 0.3 and 0.3 ± 0.8, respectively). Housing environment had an influence on tail wagging behaviour. Here, animals of the NE environment showed more tail wagging behaviour than animals of the E environment (NE 2.1 ± 3.9, E: 0.8 ± 2.4). Similarly, PES was significantly affected by housing environment (NE = 227, E = 113 and SE = 121, respectively). Treatment had a significant effect on all behavioural traits recorded with higher values in experimental animals for tail wagging (2.1 vs 0.4), play (3.4 vs 0.5) and PES (5.5 vs 0.9) than the controls. HR tended to be higher in experimental animals than in controls (p = 0.07), while RR showed a reversed tendency (p = 0.06). We conclude from these findings that novel objects can induce a positive emotional state expressed by increased play and tail wagging behaviour. High locomotor activity during the exploration of the novel objects and test arena might have superimposed on the effects of treatment and housing environment on the cardiovascular parameters. The higher tail wagging behaviour in NE animals could be interpreted as a rebound effect, reflecting the gap between their unsatisfied need for exploration in the barren home environment and the possibility to do so in the test setting. Regular confrontation with novel objects could therefore, by looking at behavioural indicators, induce positive emotional states in pigs.

Keywords

Environmental enrichment, play, tail wagging, novel object test, heart rate, heart rate variability, pigs

1. Introduction

Pigs show strong motivation to explore their surroundings in order to find resources (Spinka, 2005) and an intrinsic need to explore has been postulated (Wood-Gush and Vestergaard, 1991; Studnitz et al., 2007). Novelty is an important factor in the initiation of exploration (Berlyne, 1960) and if given a choice, pigs prefer to explore novel objects over familiar ones (Wood-Gush and Vestergaard, 1991; Moustgaard et al., 2002). In intensive pig production systems, the needs of pigs regarding cognitive challenges and the ability to express species-specific behaviour are hardly met. As a consequence, pigs kept in such production systems might predominately experience negative emotional states and their welfare is therefore considered as reduced (Murphy et al., 2014).

Emotional states are commonly characterized by a physiological (bodily reaction to sensory perception), an ethological (changes in behaviour in response to sensory perception) and a subjective component (reaction felt to sensory input; Dantzer, 1988). With regard to the physiological aspect of emotional states, heart rate (HR) and heart rate variability (HRV) have been established as potential parameters (Von Borrell at al., 2007). Concerning the ethological component, tail postures have been used as indicators of positive and negative emotional states (Kiley-Worthington, 1976; Reefmann et al., 2009). Positive emotional states may be also expressed by exploratory, consumptive or playful behaviour (Špinka et al., 2001; Paul et al., 2005; Boissy et al., 2007; Held and Špinka, 2011). In a recent study of Reimert et al. (2014) in pigs, play behaviour and tail wagging occurred significantly more often when the test pigs were given a positive treatment (access to an area filled with peat, straw and chocolate raisins). Further studies investigated how play behaviour is influenced by housing environment (Chaloupková et al., 2007, Martin et al., 2015) and showed that pigs from enriched housing environments expressed more play behaviour than their barren-housed conspecifics.

The aim of our study was to investigate the influence of different housing environments of varying enrichment on the emotional reactions of growing pigs. We assumed that the environmental enrichment affects the underlying emotional states of the pigs (Douglas et al., 2012). To elicit emotional expressions, four novel object tests were conducted. As most promising indicators of emotional states, play and tail wagging behaviour in addition to HR and HRV were individually recorded during tests.

2. Animals, Materials and Methods

2.1 Animals, housing and management

In total, 36 female piglets (Pietrain x German Landrace) from eight different litters were used as experimental animals. Born on the research farm Relliehausen of Goettingen University (Germany), experimental pigs were housed under conventional conditions. Tails were not docked. At the age of 28 days, all animals were weaned and arrived at the research facility of the Animal Science Department of Goettingen University. The animals were randomly allocated to three different housing environments so that animals of all litters were present in all three environments (Table 1).

Table 1: Description of different housing environments used in the experiment. Each pen measured 9 m².

Environment	Description
NE (non-enriched)	One third of the pen was covered with a thin layer of sawdust (approx. 0.5cm). The rest of the pen remained barren. A wooden plank (50x5x10cm) was hung up in reachable distance for the pigs as basic enrichment material.
E (enriched)	The pen was completely covered with a thick layer of sawdust (approx. 5.0cm). One third of the pen was covered with approx. 10 kg of straw. Straw was renewed once a week. A wooden plank (50x5x10cm) was provided.
SE (super-enriched)	The pen was completely covered with a thick layer of sawdust (approx. 5.0cm). One third of the pen was covered with approx. 10 kg of straw which was renewed once a week. Different enrichment materials were provided: hay, fresh grass cut, twigs with dry leaves, fresh vegetables were beaded and hung up approx. 50cm above the ground. The animals received one of the enrichments for one week (e.g. Week 1: vegetables, Week 2: twigs etc.) and within this time the enrichment was renewed every other day.

Three rooms were used, each housing the same environment. Each of the rooms (18m²) was divided in two pens (9m²) by a wooden barrier (100 cm height). In each pen, six animals were housed resulting in 1.5 m²/ animal, in total 12 animals per

74

environment. Each pen contained a subdivided lying area (about 3m²) separated by a wooden barrier (15 cm height) which the animals could easily cross. While the animals from one pen of the same environment were used as the experimental group, the animals of the other one were used as controls. The design of both pens was identical for each environment. During the whole experimental period, all animals had ad libitum access to fresh drinking water and age appropriate dry feed. From week 4 to 6 pigs were fed starter feed with 17.7% CP (Una Hakra, Hamburg, Germany, pellets), thereafter the animals received feed mixed at the research farm Relliehausen appropriate for the first fattening phase with 17.6%CP until the end of the experiment. The pens were cleaned on a daily basis. The animals were provided with daylight and additional artificial light from 07.00 – 19.00 hours. Individual weight was monitored weekly (Salter Brecknell, Smethwick, UK).

2.2 Experimental design, testing facilities and procedures

Test room

The test room consisted of three compartments (Fig. 1). All compartments were separated by solid wooden barriers. The pre-test compartment (B) was reached through a swing door from the preperation compartment (A). The testing compartment (C) was separated from the pre-test compartment by a guillotine door. Two digital video cameras were installed (Rollei Movieline SD55, Hamburg, Germany). The flooring of all compartments consisted of barren concrete ground. During testing, varying stimuli were distributed on the floor of the testing compartment (C).

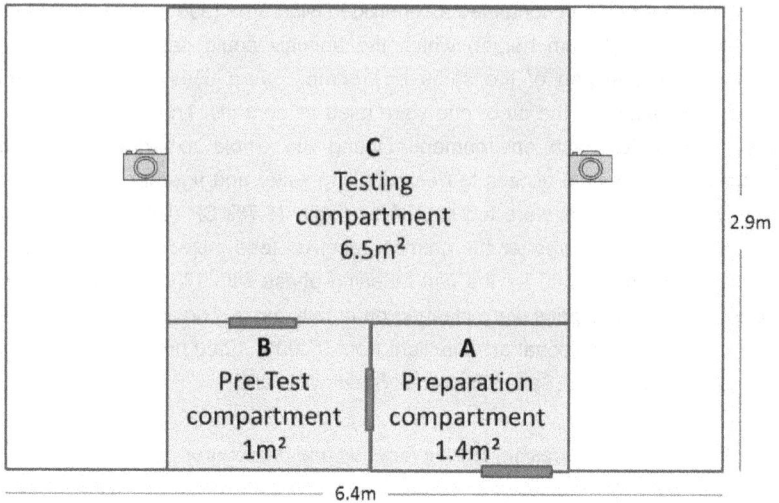

Figure 1: Test-Room layout: Pigs reached the testing area (C) via the preparation (A) and the pre-test compartment (B)

Testing procedure

After the arrival, the animals received a habituation period of seven days (animals aged 5 weeks). The experimenter sat quietly in the pens with the animals for approximately one hour daily and casually touched the animals so that they could get used to skin contact with the experimenter. During the habituation period, the animals were also introduced to the elastic belt which was later used for measuring heart rate and HRV. All animals were taken to the test room twice to habituate to the test surroundings as well.

Tests were conducted in postnatal weeks 6 to 7. All animals were transferred to the test room which was considered as an open field (Murphy et al., 2014). For experimental pigs (N=18) the open field was enriched with novel objects, whereas control animals (N=18) found the barren open field only. Animals from each pen were subdivided into fixed groups of three animals each, resulting in two experimental and two control groups for each environment. During the entire study, all tests were conducted in the same groups to avoid social isolation stress. Each animal passed 4 test runs with 1 day between tests. In total, 144 tests were conducted. Test duration was 7 minutes.

During testing, behaviour was continuously recorded by two video cameras (Rollei Movieline SD55, Hamburg, Germany) with 30 frames/ sec. For measurement of vascular responses the Polar® RS800CX (Electro Oy, Finland) was used. Both integrated electrodes of an elastic belt were coated with electrode gel to enhance conductibility. A sending unit was connected to the elastic belt, transferring the HR data to a receiving unit. Both heart beats and the intervals between heartbeats (RR-Intervals) were recorded. The Polar® system was validated for the use in pigs by Marchant-Forde et al. (2004) and was also used for behaviour tests in young pigs by Zebunke et al. (2011).

For the test procedure, all animals were individually marked by a number sprayed on their back. Tests were carried out on 8 test days in the morning (9:00 – 11:00 h) and the afternoon (13:00 – 15:00 h). The sequence of the test groups was randomized across daytime and housing environment. For testing, the doors of the home pen were opened and all pigs were allowed to proceed into the hallway. From there, test individuals were led into the preparation compartment, while the remaining pigs returned from the hallway to the home pen.

In the preparation compartment the elastic belts of the Polar® system were attached around the thorax of the test animals, directly behind the front legs and the electrodes were placed across the sternum. Timeframe for these procedures was set to one minute for all three individuals. The animals were then led into the pre-test compartment. The experimenter left the test room and the animals remained in the pre-test compartment for one minute. The experimenter re-entered the test room and led the animals further into the testing compartment. Sows from the experimental groups were confronted with one of the four different novel objects and materials (a wallow, soil, rubber dog toys filled with grapes, rubber ducks, Table 2). These objects were chosen based on their attractiveness in a previous study (Chapter 2). Three of the same objects were laid out on the test room floor (one for each animal). The wallow was presented as one and the soil was presented in a heap. In contrast, control animals were exposed to the barren testing area. The experimenter left the test room and re-entered after seven minutes. Pigs were led back through the pre-test into the preparation compartment. Elastic belts were taken off and the animals were moved back to their home pen.

Table 2: Novel Objects used in the behaviour tests

Novel Objects	Description
Soil	A heap of potting soil (50l) was placed in the center of the testing arena.
Wallow	A wooden frame (163 X 90 cm) was laid out with pondliner and filled with potting soil and water. This wallow was placed in the center of the testing arena.
Ducks	Three yellow rubber ducks (9 cm height) which squeaked when pressed were laid out on three marked spots in the testing arena
Kong®	Three red rubber dog toys "Kong"® (7 cm height) were filled with grapes and laid out on three marked spots in the testing arena.

2.3 Behavioural Analysis

Video recordings made during the tests were individually analyzed for the frequency of play bouts and tail wagging behaviour shown by animals of experimental and control groups by continuous behavioural sampling (Martin and Bateson, 2007) and expressed as the total frequency of occurrence. Tail wagging was defined by the pigs' tail swinging in any direction. Play behaviour was defined as running, gamboling, pivoting, carrying or throwing objects around with the animals' snout.

2.4 Heart Rate Analysis

HR and HVR recordings during which the connectivity between sending and receiving unit was lost during the testing time, were discarded from further analysis. Only uninterrupted segments of five min (Sloan et al., 1994; Von Borrell et al., 2007), starting when the animals entered compartment C, were included. Thus, only 31 (22%) of the heart rate measurements were suitable for analysis. All remaining recordings were analysed using the Kubios Software for HRV Analysis (Biosignal and Medical Imaging Group,University of Kuopio, Department of Applied Physics, Kuopio, Finland, for review see Niskanen et al., 2002) applying the software's medium artefact correction level. Heart rate (beats per minute, bpm) and heart rate variability was analysed using the parameter RR (duration of intervals between consecutive heart beats; see von Borell et al., 2007) and RMSSD (root mean square of the difference between successive RR intervals), reflecting vagal activity only.

2.5 Statistical Analysis

Statistical analyses were performed using SAS (SAS 9.3, SAS Institute Inc.) and SPSS (IBM SPSS Statistics 24). Tail wagging and play only occurred at low frequencies. Therefore, individual frequencies of play and tail wagging behaviour were also summarized to form the Positive Emotion Score (PES). For the analysis of behavioural data underlying a Poisson distribution (play, tail wagging and PES), a generalized linear model using the GLIMMIX procedure was used. Fixed effects of the model were environment, treatment, object nested within treatment and the interactions thereof. Animal was considered random. HR, RR and RMSSD were normally distributed and the same general mixed model, using the GLIMMIX procedure was used.

For a logistic regression analysis, PES was converted into a binary variable and its relation with the cardiovascular parameters was calculated using SPSS (IBM SPSS Statistics 24). Statistical significance was set at $P < 0.05$.

3. Results

3.1 Behavioural traits

The frequencies of tail wagging and play were low as shown in Tab. 3 and Fig. 2. Housing environment had no effect on play (p = 0.14). In contrast, tail wagging behaviour differed significantly among housing environments (NE = 2.1, E= 0.8 and SE = 0.8, respectively) with significantly more tail wagging in NE animals than in animals from the E environment (p = 0.03). Similarly, PES was significantly affected by housing environment (NE = 227, E = 113 and SE = 121, respectively). Treatment had a significant effect on all behavioural traits recorded with higher values in experimental animals for tail wagging (2.1 vs 0.4), play (3.4 vs 0.5) and PES (5.5 vs 0.9) than the controls. The effect of the different novel objects was tested for the experimental animals and was found to be significant for tail wagging, play behaviour and PES (p = 0.0001; Tab. 3). Highest values were found for soil and wallow, followed by the Kong® while exposure to the rubber ducks resulted in the lowest reactions.

Significant environment x treatment interactions were found for play behaviour and PES (Fig. 2). Within each environment, the experimental animals had a higher PES than the controls.

With regard to play behaviour, the experimental animals showed more play behaviour than the controls in both the NE and the E environments, but not in the SE housing (NE: experimental 4.8 ± 6.7 vs control 0.4 ± 1.0, E: experimental 3.0 ± 3.2 vs control 0.1 ± 0.3).

Table 3: Frequencies of play, tail wagging behaviour and PES (LSQmeans ± SEM) for experimental (across novel objects) and control animals (across all tests).

Trait	Experimental animals				Control Animals[1]	Significance of effects			
	Novel object					P-values			
	Kong®	Soil	Wallow	Rubber ducks		Environment	Treatment	Object (Treatment)	Environment* Treatment
Tail Wagging	1.4 ± 0.9^{aA}	3.3 ± 1.2^{aB}	2.9 ± 1.2^{aB}	0.7 ± 0.3^{aA}	0.0 ± 0.8^{b}	0.03	0.001	0.0001	0.09
Play	2.6 ± 0.5^{aA}	3.9 ± 1.5^{aA}	6.7 ± 1.2^{aB}	0.5 ± 0.3^{aC}	0.0 ± 1.2^{b}	0.14	0.0001	0.0001	0.03
PES	4.0 ± 1.2^{aAB}	7.2 ± 2.5^{aAB}	9.6 ± 1.9^{aA}	1.2 ± 0.4^{aB}	0.0 ± 0.8^{b}	0.04	0.0001	0.0001	0.01

[1] Across all tests. Control animals were tested in the same open field as the experimental pigs, but without novel objects.
a,b Diffrerences between treatments, p <0.05
A,B Differences between novel objects, p <0.05

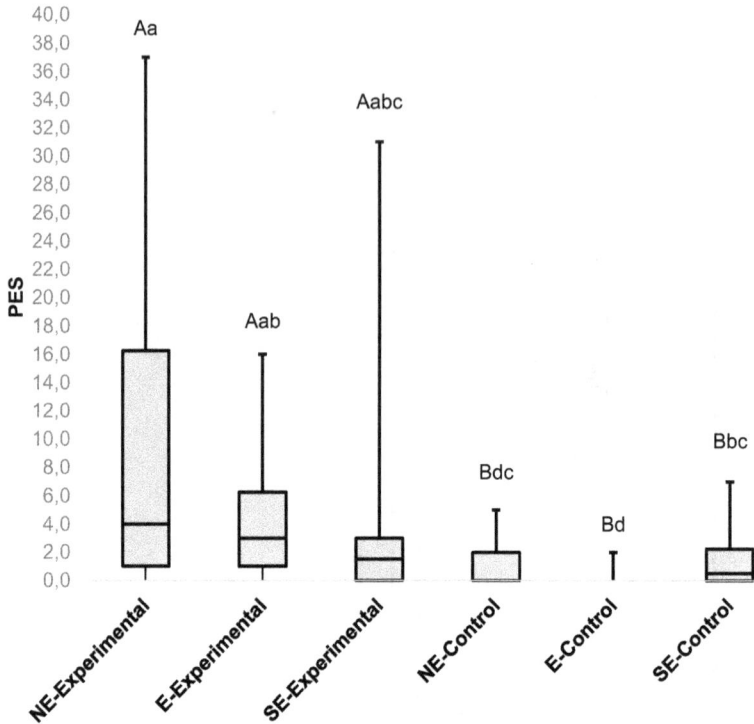

Figure 2: Frequencies of play and tail wagging bouts summarized as positive emotion score (PES) in pigs (n=36) across different treatments (experimental and control) and housing environments with varying enrichment levels (NE=non-enriched, E=enriched, SE=super-enriched). Different small letters indicate significant differences in PES between housing environments, capital letters those between treatments.

3.2 HR, HRV and RMSSD

Data on HR, RR and RMSSD for all environments and treatments are presented in Table 4. Environment had no effect on either of the parameters. HR tended to be higher in experimental animals than in controls (180.1 vs 171.4, p=0.07), while RR showed a reversed tendency (337.4 vs 353.9, P=0.06). Treatment had no significant effect on RMSSD. The environment x treatment interactions were not significant either.

The logistic regression analysis resulted in a non-significant model (Chi-Square (3) = 4.89, P = 0.77, N=31).

Table 4: Heart rate (HR) and heart rate variability (HVR) parameters (LSmeans ± SEM) measured in pigs during behavioural tests, by housing environment, and treatment. Varying N numbers result from varying numbers of usable recordings.

Traits	Housing						Significance of effects p-values			
	Non-Enriched		Enriched		Super-Enriched					
	Experimental (N=6)	Control (N=5)	Experimental (N=7)	Control (N=6)	Experimental (N=4)	Control (N=3)	Environ-ment	Treat-ment	Object (Treatment)	Environment* Treatment
HR	180.2 ± 7.8	161.9 ± 4.6	175.8 ± 4.8	182.7 ± 5.2	187.6 ± 5.1	164.5 ± 2.4	0.39	0.07	0.23	0.09
RR	334.2 ± 12.5	374.0 ± 11.0	345.4 ± 9.4	334.0 ± 11.6	321.5 ± 8.8	366.5 ± 4.8	0.28	0.06	0.29	0.08
RMSSD	11.3 ± 1.3	12.8 ± 1.2	9.9 ± 1.2	7.8 ± 1.2	7.7 ± 1.3	9.2 ± 2.0	0.10	0.97	0.16	0.38

4. Discussion

We investigated the influence of different housing environments of varying enrichment on the emotional reactions of growing pigs. We hypothesized that housing environment would have a significant effect on the underlying emotional states of the pigs. Novel object tests were used to elicit emotional expressions, measured by cardiovascular and behavioural parameters. The pigs' behavioural reactions in the test situation showed clear differences between housing environments. The highest reactions were found in experimental animals from the most impoverished environment (NE) when confronted with novel objects, while the housing effect was absent in control animals. This significant housing environment x treatment interaction underlines the effectiveness of the chosen experimental approach. Thus, in the novel object tests more emotional expressions were elicited and a possible underlying motivational bias was enhanced. Accordingly, the expected differentiation between housing environments became more pronounced.

Behavioural reactions, such as tail postures have been used as indicators of positive and negative emotional states (Kiley-Worthington, 1976; Reefmann et al., 2009). Positive emotional states can be expressed by exploratory, consumptive or playful behaviour (Špinka et al., 2001; Paul et al., 2005; Boissy et al., 2007; Held and Špinka, 2011). In a study of Reimert et al. (2014) in pigs, play behaviour and tail wagging occurred significantly more often when the test pigs were given a positive treatment (access to an area filled with peat, straw and chocolate raisins) than when given a negative treatment (isolation and restricted movement). Similarly, the increase in tail wagging behaviour in our pigs during the novel object tests can be interpreted as indicative of a positive emotional state in the experimental animals.

According to the theory of appraisal, positive emotional states are triggered by events or stimuli that do not occur suddenly, are fairly predictable and familiar to an individual and entail a degree of pleasantness (studied by Sander et al., 2005 in humans and Veissier et al., 2009 in sheep). Our findings support these theories firstly because the novel objects in our test were laid out in the arena already before the animals entered, so they did not occur suddenly. Secondly, the animals were habituated to the test room beforehand, so the situation was fairly predictable and familiar to them, and a fixed group of conspecifics provided social support. Lastly, the objects were chosen on the basis of pleasantness for the animals (substrate for rooting, food rewards), supporting the assumption that the experimental animals experienced more positive emotions than the controls.

Play behaviour cannot only be observed in the context of positive emotion, as animals also showed play behaviour when coping with negative situations (Murphy et al., 2014). For instance, pigs housed in enriched environments exhibited more play behaviour than barren housed pigs in their home pen (Bolhuis et al., 2005, Chaloupková et al., 2007). When moved to a novel pen, however, barren housed pigs were observed to display more play behaviour than pigs from enriched environments (Wood-Gush et al., 1990). Barren housing environments therefore facilitate play behaviour less and the lack of stimulation might increase the animals' motivation for, e.g. exploratory behaviours. As soon as the situation changes and the animal is given the opportunity to perform said behavioural pattern, the frequency or speed with which the behaviour will be shown will be larger or faster (Lorenz, 1950). Such so-called rebound effects (Kennedy, 1985) have been observed in various animal species (hens: Nicol, 1987; horses: McGreevy and Nicol, 1998; rabbits: Dixon et al., 2010). In our study, housing environment affected tail wagging behaviour and the derived PES with animals of the barren NE housing showing more tail wagging behaviour. It is suggested, that animals of the barren environment were not able to satisfy their need for exploration to the same extent as their enriched conspecifics and therefore might have experienced an increased positive affective state during novel object tests. Their increased tail wagging during testing can be interpreted as a rebound effect (Kennedy, 1985).

The novel objects used in this study elicited significant differences in play, tail wagging and the PES in the experimental animals. Highest values were found for soil and wallow, followed by the Kong® while exposure to the rubber ducks resulted in the lowest reactions. This ranking in attractiveness agrees with previous results (Chapter 2) and underlines earlier suggestions that attractive materials for pigs are characterized by being changeable, destructible as well as edible (Studnitz et al., 2007). The Kong®, which was food-baited, also resulted in play behaviour where the food bait might have acted as a trigger. Searching for food is an essential behaviour in pigs, so that they are even choosing to root for hidden food, instead of eating equal food provided from a trough (De Jonge, 2008). We have found similar object preferences across all environments. So, even though the animals of the enriched environments E and SE had large amounts of substrate available in their home environments at any time, in the test situation they still preferred the substrated novel objects (soil and wallow) over the Kong® and the rubber ducks. This finding might add to the importance of exploring substrate in pigs.

Contrary to our expectations from the behavioural differences found between environments, the cardiovascular parameters were not significantly affected by housing environment or treatment. This raises questions about the physiological indicators of possible emotional states of the pigs during the tests. According to various studies, emotional states and therefore subjective experiences of situations can be assessed in terms of two underlying dimensions. First, it is important to know whether a situation is perceived as positive vs negative (or pleasant vs unpleasant), referring to its *valence* for the individual. The second dimension refers to the associated level of *arousal* the situation causes in the individual (high vs low) (Russell and Barrett, 1999; Burgdorf and Panksepp, 2007; Mendl et al., 2010). Whereas situations of positive valence and low arousal refer to calm and relaxed states of contentment (Carver, 2001), positively valenced states of high arousal are considered to reflect excitement and happiness and are associated to appetitive motivational states or seeking behaviours and occur upon the signalling or presentation of rewards (Rolls, 2005; Burgdorf and Panksepp, 2006; Mendl et al., 2010). The confrontation with the test situation in our study was therefore probably a situation of both positive valence and high arousal for the experimental pigs. A state of high arousal could explain the lack of differences in cardiovascular parameters in both, the treatment and the control pigs across all housing environments. The different underlying dimensions might also explain the non-significant logistic regression between behavioural and cardiovascular parameters. However, the small number of usable cardiovascular measurements limits further conclusions.

Treatments such as the novel object test for the experimental animals and the open-field-test for the control animals in our study induce increased behavioural and locomotor reactions which may not be seen under controlled conditions, e.g.in the animals' home environment (Von Borell et al., 2007). All animals of our study were given the opportunity to explore the test room twice prior to the actual novel object test. However, habituation to test conditions prior to the behaviour test might not have been sufficient as suggested by Murphy et al. (2014). Thus, the test situation may cause anxiety in the animals during the novel object test, masking positive emotional states that might be induced by the test. This context might also explain the incoherence between the physiological and ethological observations in this study.

During novel object tests, pigs showed a higher degree of tail wagging behaviour which is suggested as indicator of a positive emotional state. Accordingly, the present novel object tests were positively valenced by the animals. However, high arousal and increased locomotor activity through exploration might result in high HR and low HRV, which limits a conclusive interpretation of the underlying emotional states based on cardiovascular parameters. Further research, especially comparing cardiac parameters during the test situation with the animals' basal levels in the home pen, could further elucidate the impact of housing environment on emotional states in pigs. The observed rebound effect in NE animals during novel object tests highlights the importance to attain possible positive emotional states. On farm, regular confrontation with novel objects could be used to elicit positive emotional states in pigs and therefore increase their well-being.

References

Berlyne, D. E. 1960. Conflict, arousal, and curiosity. McGraw-Hill, New York.

Boissy, A., Manteuffel, G., Jensen, M.B., Moe, R.O., Spruijt, B., Keeling, L.J. 2007. Assessment of positive emotions in animals to improve their welfare. Physiol. Behav. 92, 375–397.

Burgdorf, J., Panksepp, J. 2006. The neurobiology of positive emotions. Neuroscience & Biobehavioral Reviews, 30, 173-187.

Carver, C. S. 2001. Affect and the functional bases of behavior: On the dimensional structure of affective experience. J. Pers. Soc. Psychol. Rev. 5, 345-356.

Chaloupková, H., Illmann, G., Bartoš, L., & Špinka, M. 2007. The effect of pre-weaning housing on the play and agonistic behaviour of domestic pigs. Appl. Anim. Behav. Sci. 103, 25-34.

Dantzer, R. 1988. Les Émotions, Presses Universitaires de France, Collection Que sais-je?, Paris, 121 pp.

Dawkins, M. 1983. La Souffrance Animale. Editions du Point Vétérinaire, Maisons-Alfort 152.

Douglas, C., Bateson, M., Walsh, C., Bédué, A., Edwards, S. 2012. Environmental enrichment induces optimistic cognitive biases in pigs. Appl. Anim. Behav. Sci. 139, 65-73.

Fraser, D. (1995) Science, values and animal welfare: exploring the 'inextricable connection'. Anim. Welf. 4, 103-117.

Held, S. D., Špinka, M. 2011. Animal play and animal welfare. Anim. Behav. 81, 891-899.

Marchant-Forde, R. M., Marlin, D. J., Marchant-Forde, J. N. 2004. Validation of a cardiac monitor for measuring heart rate variability in adult female pigs: accuracy, artefacts and editing. Physiol. Behav. 80, 449-458.

Martin, J. E., Ison, S. H., Baxter, E. M. 2015. The influence of neonatal environment on piglet play behaviour and post-weaning social and cognitive development. Appl. Anim. Behav. Sci., 163, 69-79.

Mendl, M., Burman, O. H., Paul, E. S. 2010. An integrative and functional framework for the study of animal emotion and mood. Proc. R. Soc. London B: Biol. Sci., 277, 2895-2904.

Moustgaard, A., Lind, N.M., Hemmingsen, R., Hansen, A.K., 2002. Spontaneous object recognition in the Göttingen Minipig. Neur. Plast. 9, 255–259.

Murphy, E., Nordquist, R.E., Van der Straay, F.J. 2014. A review of methods to study emotion and mood in pigs, *Sus scrofa*. Appl. Anim. Behav. Sci. 159, 9-28.

Paul, E.S., Harding, E.J., Mendl, M. (2005) Measuring emotional processes in animals: the utility of a cognitive approach. Neurosci. Biobehav. Rev. 29, 469-491.

Panksepp, J. 1998. Affective Neuroscience: The Foundations of Human and Animal Emotion. Oxford: Oxford University Press

Reimert, I., Bolhuis, E.J., Kemp, B., Rodenburg, B. 2015. Emotions on the loose: emotional contagion and the role of oxytocin in pigs. Anim. Cogn. 18, 517-532.

Rolls, E. T. 2005. Emotion explained. Oxford University Press, USA.

Russell, J. A., & Barrett, L. F. 1999. Core affect, prototypical emotional episodes, and other things called emotion: dissecting the elephant. J. Pers. Soc. Psychol. 76, 805.

Sander, D., Grandjean, D., Scherer, K.R. 2005. A systems approach to appraisal mechanisms in emotion: emotion and brain. Neur. Netw. 18, 317-352.

Špinka, M., Newburry, R.C., Bekoff, M. 2001. Mammalian play: training for the unexpected. Quart. Rev. Biol. 76, 141-168.

Špinka, M. 2009. Behaviour of pigs. In: The Ethology of Domestic Animals: An Introductory Text (p. 177). CAB International, Wallingford.

Studnitz, M., Jensen, M. B., Pedersen, L. J. 2007. Why do pigs root and in what will they root?: A review on the exploratory behaviour of pigs in relation to environmental enrichment. Appl. Anim. Behav. Sci. 107, 183-197.

Vanderschuren, L. J. 2010. How the Brain Makes Play Fun. American J. Play. 2, 315-337.

Veissier, I., Boissy, A., Désiré, L., Greiveldinger, L. 2009. Animals' emotions: studies in sheep using appraisal theories. Anim. Welf. 18, 347-354.

Wood-Gush, D. G. M., Jensen, P., & Algers, B. 1990. Behaviour of pigs in a novel semi-natural environment. Biol. Behav. 15, 62-73.

Wood-Gush, D. G. M., Vestergaard, K. 1991. The seeking of novelty and its relation to play. Anim. Behav. 42, 599-606.

Zebunke, M., Puppe, B., Langbein, J. 2013. Effects of cognitive enrichment on behavioural and physiological reactions of pigs. Physiol. Behav. 118, 70-77.

CHAPTER IV

The influence of housing enrichment and novelty on cognitive abilities in
juvenile pigs (*Sus scrofa*)

L. Mckenna[1] and M. Gerken[1]

[1]Department of Animal Sciences, University of Goettingen, Albrecht-Thaer-Weg 3,
37075 Goettingen, Germany

Animal Welfare

Submitted April 2018

Abstract

Pigs kept in intensive husbandry systems undergo various limitations with regards to the expression of species specific behaviour. These limiting living environments can lead to behavioural disorders and influence cognitive abilities of pigs. The aim of this study was to investigate the problem solving abilities of 36 female pigs (aged 7-8 weeks), housed in environments with different levels of enrichment (non-enriched (NE), enriched (E), super-enriched (SE)) by means of two cognitive tasks (board and pipe). The animals in each environment were divided into test and control groups: prior to the cognitive tasks, the test animals had undergone a series of novel object tests, whereas the control animals found the test room empty. For the present behavioural testing, all animals were led from their home environment to the test arena where they were confronted with a baited board or a pipe. The tasks were considered solved when the pigs successfully retrieved a hidden food item. No differences were found between environments with regards to the number of solved and unsolved tasks for board and pipe task ($p = 0.23$, $p = 0.97$, respectively). In both tests animals of the NE environment tended to approach the test apparatus and to solve the task faster than the animals of the E environment ($p = 0.07$ and $p = 0.09$). The test animals solved the board task faster than the control animals ($p = 0.03$), while similar differences were not apparent in the pipe task. These results underline the importance of novelty experience in cognitive tasks. We conclude that the pigs kept in non-enriched environments have retained a high motivation to explore and interact with stimuli and were therefore faster to touch and solve the cognitive tasks. Cognitive enrichment is an important tool to create situations which allow pigs to have a certain degree of control over their environment combined with the opportunity to gain a reward.

Keywords: animal welfare, cognition, enrichment, environment, motivation, pig

1. Introduction

Most modern intensive farming systems used for the production of pig meat can be characterized as barren and allow only little space. These stimulus-poor environments limit the development and performance of normal pig behaviour (Bolhuis et al., 2013).

Commercial environments are in contrast to the natural habitat of the domestic pigs' ancestor, the wild boar, which is highly diverse, ranging from coastal scrubland to mountainous areas as well as farmland, swamps or forests (Frädrich, 1974). Wild boar and feral pigs do not show major differences regarding their behaviour (D'Eath and Turner, 2009). Taking the behaviour of the domestic pig kept in extensive conditions into comparison, behavioural differences are also more of a quantitative and less of a qualitative nature (D'Eath and Turner, 2009). In essence, both wild and domestic pigs have shown to be inquisitive, highly curious, social animals who spend around 75% of their waking hours actively exploring their environment (Stolba and Wood Gush, 1989).

Pigs have shown substantial cognitive abilities which are necessary for example to locate and memorise feeding sites (Held et al., 2010, Grimberg-Henrici et al., 2015, Van der Staay et al., 2016). Mendl et al. (1997a) showed that pigs searching for a hidden food item in one of ten enclosed areas used their memory to relocate the hidden food and avoided re-visits to empty areas in later trials. Pigs performed above chance, when discriminating baited from un-baited pots based on visual and on olfactory cues in a novel environment (Croney et al., 2003). Furthermore, pigs have shown learning ability through operant conditioning (Sneddon et al., 2000, Ferguson et al., 2009), spatial, visual, olfactory and social discrimination (Croney et al., 2003, McLeman et al., 2005). Also, observation learning has been described in pigs by means of a mirror test (Broom et al., 2009). Göttingen minipigs demonstrated that they can solve the spatial DNMS task, a well-validated memory test (Nielsen et al. 2009). In the study of Lind and Moustgaard (2005), minipigs learned a Go/No-go task and it was underlined that emotional factors may influence their learning rate.

In early work on the influence of different environments it was shown that rats housed under enriched conditions develop differences in brain measures (higher enzymatic activity in the cortex) and aspects of their behaviour (better learning and problem-solving abilities) compared to conspecifics kept under impoverished conditions (Rosenzweig et al., 1964, Rosenzweig & Bennet, 1996). The possible influence of the housing environments on pigs' cognitive abilities was focussed by several studies (Mendl et al., 1997b). Pigs housed in substrate enriched environments were compared with those living in substrate impoverished environments in a spatial task (a t-maze). The substrate impoverished pigs spent more time investigating the maze, whereas substrate enriched pigs focused on finding the food source and quickly moved towards it. Further, the substrate enriched animals were less able to change their behaviour when the food source was located at the other arm of the T-maze (Mendl et al., 1997b). In the study of Grimberg-Henrici et al. (2015) pigs housed in barren or enriched environments had to learn a fixed pattern of baited and un-baited holes in a holeboard task and solve the same task in a reversal test with a diagonally mirrored image. The pigs from the enriched environment took less time to search for the food items, exhibited reduced visits to the un-baited holes and also showed reduced re-visits of baited and un-baited holes which they had already visited compared to the barren housed pigs.

Brain plasticity induced by differing living environments (enriched, non-enriched or isolated) has been shown to occur over the whole life span (Rosenzweig et al., 1964, Rosenzweig & Bennet, 1996). However, influences on the brain occur more rapidly in younger animals and the effect these changes have on the young animal is larger than in the older animal (Rosenzweig & Bennet, 1996) as shown in e.g. squirrels (Renner and Rosenzweig, 1987). Therefore, the effect of varying stimulation of housing environments on the plasticity of the brain in young pigs might play an important role when investigating their cognitive abilities.

The aim of the present study was to investigate the effect of three different living environments (non-enriched (NE), enriched (E) and super-enriched (SE)) on the ability of juvenile pigs to solve instrumental tasks by retrieving a food item. Half of the test animals from each group had the opportunity of exploring various novel objects and materials prior to the cognitive tasks. Therefore, we also investigated whether previous stimulation had an additional effect on the ability of the test pigs to solve the tasks and retrieve the food item.

2. Animals, Materials and Methods

2.1 Animals, housing and management

In total, 36 female piglets (Pietrain x German Landrace) from eight different litters were used as experimental animals. Born on the research farm Relliehausen of Goettingen University (Germany), experimental pigs were housed under conventional conditions until weaning. Tails were not docked.

At the age of 28 days, all animals arrived at the research facility of the Animal Science Department of Goettingen University. The animals were randomly allocated to three different housing environments so that animals of all litters were present in all three environments (Table 1).

Table 1: Description of different housing environments used in the experiment.

Environment	Description
NE (non-enriched)	One third of the pen was covered with a thin layer of sawdust (approx. 0.5 cm). The rest of the pen remained barren. In order to fulfil legal requirements for pig husbandry (EU directive 2008/120/EC), a wooden plank (50x5x10 cm) was hung up in reachable distance for the pigs as basic enrichment material.
E (enriched)	The pen was completely covered with a thick layer of sawdust (approx. 5.0 cm). One third of the pen was covered with approx. 10 kg of straw. Straw was renewed once a week. A wooden plank (50x5x10 cm) was provided.
SE (super-enriched)	The pen was completely covered with a thick layer of sawdust (approx. 5.0cm). One third of the pen was covered with approx. 10 kg of straw which was renewed once a week. Different enrichment materials were provided: hay, fresh grass cut, twigs with dry leaves, fresh vegetables were beaded and hung up approx. 50 cm above the ground and wooden logs were hung up approx. 50 cm above the ground. The animals received one of the enrichments every week (e.g. Week 1: vegetables, Week 2: twigs etc.) and within this time the enrichment was renewed every other day.

Each of the three environments (18 m²) was divided in two pens (9 m²) of the same size by a wooden barrier. In each compartment, six animals were housed with 1.5 m²/ animal, resulting in a total of 12 animals per environment. Each pen was subdivided into a lying area (about 3 m²) separated by a wooden slat. The design of both compartments was identical for each environment. While the animals from one compartment were used as the test group, the animals of the other one were used as controls.

During the whole experimental period, all animals had ad libitum access to fresh drinking water and age appropriate concentrate (Una Hakra, Hamburg, Germany, pellets): starter feed with 17.7% CP from week 4 to 6, thereafter the animals received feed mixed at the research farm Relliehausen appropriate for the first fattening phase with 17.6%CP until the end of the experiment. The pens were cleaned on a daily basis. Fresh sawdust or straw was added when necessary. The animals were provided with daylight and additional artificial light from 07:00 to 19:00 hours. The weight of all animals was monitored weekly to the nearest 0.1g (Salter Brecknell, Smethwick, UK).

2.2. Experimental design, testing facilities and procedures

The experiment consisted of three phases. During the habituation period of seven days (week 5), the animals had the opportunity to habituate to the new surroundings, their new pen mates and the experimenter. Therefore, the experimenter sat quietly in the experimental pens with the animals for approximately one hour daily and casually touched the animals so that they could get used to skin contact with the experimenter. During the habituation period all animals were taken to the test room twice in order to get used to the test surroundings. Animals from each pen were subdivided into two groups of 3 animals each, resulting in 4 groups per environment. During the entire experiment, all tests were conducted in the same groups to avoid social isolation stress.

The test room consisted of three compartments (Figure 1) placed on the concrete floor. All compartments were separated by wooden barriers. The animals were led in their same test group from their home pen over a hallway through the main door into the preparation compartment. The pre-test compartment was reached through a swing door from the preparation compartment whereas the animals had to go through a guillotine door from the pre-test compartment to the testing compartment. The doors were opened by the experimenter with a cable pull for the pigs to pass through. Behaviour of the animals was continuously recorded by two digital video cameras (Rollei Movieline SD55, Hamburg, Germany, 30 fps, 1920x1080p).

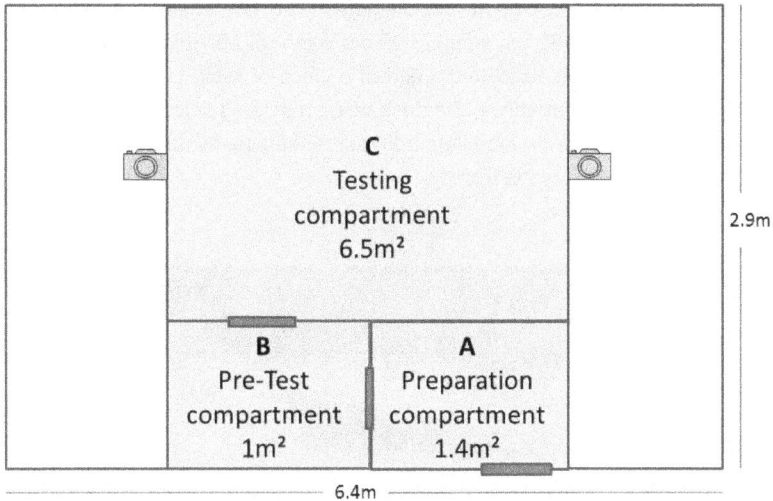

Figure 1: Test-Room layout: Pigs reached the testing area (C) via the preparation (A) and the pre-test (B) compartment.

In experimental phase 2 (weeks 6 to 7), all animals were repeatedly moved in their groups to the test room and were released to the testing area for 7 min. Sows from the test treatment were confronted with different novel objects and materials (a wallow, soil, rubber dog toys, rubber ducks). Three of the objects each were laid out on the test room floor (one for each animal). The wallow was presented as one and the soil was presented in a heap. Control animals were exposed to the barren testing area. After the test, animals were returned to their home pen. Each animal passed 4 test runs with 1 day between tests.

In experimental phase 3 (week 8 to 9), two tasks were presented to the pigs in order to evaluate their problem solving ability:

1. "Board": A wooden board (37cm length x 29 cm width) was angled at 45° (Figure 2). A food item (grape) was then attached to the end of piece of twine (1 m length), and placed underneath the angled board. The task was solved once the animal managed to retrieve the food item from under the board by pulling the rope or tilting and pushing away the board.

2. "Pipe": A plastic pipe (5 cm diameter, 31.5 cm length) was attached to a wooden board (37 cm length x 29 cm width) at 90° (Figure 3). Again, a food item (grape) was fixed to the end of a piece of twine (1m length) and placed into the pipe from above. The task was considered solved as soon as the test animal retrieved the food item from out of the pipe by pulling the rope or tilting and pushing away the board.

Figure 2: Cognitive Task "Board": A grape attached to a rope and placed under an angled board

Figure 3: Cognitive Task "Pipe": A grape attached to a rope and hung into the pipe

All animals entered the test room in groups of three where they were led via the preparation compartment into the pre-test compartment. There, the animals remained for one minute while the experimenter left the test room. After re-entering, the experimenter led one of the three test animals into the test arena, where the task (board or pipe) was already positioned. The other two test animals remained in the pre-test compartment. The experimenter left the test room and the animal in the test arena had one minute to solve the task. After one minute, the experimenter re-entered the test room, led the test animal back into the pre-test compartment, re-stocked the task and positioned it back to the designated place in the test arena. Thereafter, the second test animal was led into the test arena in order to solve the task. The same procedure was performed with the third animal. The order of the animals entering the test arena was randomized. All animals were first tested twice on the board task and then underwent two tests on the pipe task. The interval between tests was 1 day for the individual.

2.3 Behavioural Analysis

The test runs of all animals were video recorded and individually analysed for the following traits:

- Solving task: the solving success was classified into three categories

  (i) retrieving food item by pulling the rope

  (ii) alternative strategy: retrieving food item by pushing away or tilting the test apparatus)

  (iii) not retrieving food item within maximum test duration of 60 sec

- Latency First Touch: time from entering the test arena to touching the test apparatus for the first time (s)
- Latency Solved: time from entering the arena to retrieving food item (s)
- Latency First Touch – Solved: the time from first touching the test apparatus until retrieving food item (s)

When the task was not solved within the maximum limit of 60 sec, all latencies were set to 60 sec.

2.4 Statistical analyses

Statistical analyses were performed using SAS (SAS 9.3, SAS Institute Inc.) for each test separately. The mixed linear model included treatment, environment, group within environment, trial and the respective interactions as fixed effects and the animal as random effect. Residuals were checked for normality. If the normality assumption was not met, square root transformations were used. The effect of 'group' was not significant and was then omitted from the final model. Mixed linear model analysis was applied to the variables "Latency Solved" and "Latency First

Touch-Solved" and included only the values of the animals that solved the task within the given time of 60s. For further analyses solving categories were also converted into solved (both solution strategies combined) and unsolved.

For both cognitive tasks (board and pipe), normal distribution of the variable 'Latency First Touch' could not be realized. This variable was therefore analysed with the Mann Whitney U Test including data of all animals. For sows that did not touch the test apparatus within the whole duration of the test the maximum time of 60s was allotted. The number of animals solving the task was analysed with the Chi² Test. Data are expressed as means ± SEM. Statistical significance was set at $P < 0.05$.

3. Results

3.1 Cognitive Task: Board

The number of solved, alternatively solved and unsolved tasks is presented in Tables 2 and 3. Housing environment neither influenced the first nor the second trial with regards to the proportion of solved trials (Chi² test, chi² = 5.56, p = 0.23). Non-enriched sows tended (p = 0.07) to touch the test apparatus and to solve the task faster than sows of the other environments (Table 4 and 5). Test animals which had undergone previous novel object exposure required significantly less time (latency solved) than the control animals (p = 0.03). Interestingly, the treatment x environment interaction was not significant. In the second trial, all sows tended to show more rapid solving capacities than in the first test.

Table 2: Number (N) and Proportion (% in parenthesis) of solving results: solved (retrieving food by pulling rope), alternatively solved (retrieving food by tilting or pushing away test apparatus) and unsolved trials (not retrieving food) for both cognitive tasks (board and pipe) by environment (NE = non-enriched, E = enriched, SE = super-enriched) and across two trials per test

Cognitive Test	Solution category	Housing environment			Total (%)
Board		NE	E	SE	
	Solved	12 (50)	5 (20.8)	6 (25)	31.9
	Alternatively solved	8 (33.3)	12 (50)	12 (50)	44.4
	Unsolved	4 (16.7)	7 (29.1)	6 (25)	23.6
Pipe					
	Solved	12 (50)	12 (50)	12 (50)	50.0
	Alternatively solved	1 (4.2)	2 (8.3)	2 (8.3)	6.9
	Unsolved	11 (45.8)	10(41.7)	10 (41.7)	43.1

Table 3: Number (N) and proportions (% in parenthesis) of solved trials[1] for test (n=18) and control (n=18) animals for both cognitive tests (Board and Pipe) by trial number and housing environment.

Cognitive Test	Housing environment		
Board	Non-Enriched	Enriched	Super-Enriched
First Trial (%)	10 (83.3)	7 (58.3)	8 (66.7)
Second Trial (%)	10 (83.3)	10 (83.3)	10 (83.3)
Pipe			
First Trial (%)	5 (41.6)	6 (50.0)	7 (58.3)
Second Trial (%)	8 (66.7)	8 (66.7)	7 (58.3)

[1] Solved: solution categories 'solved by pulling rope' and 'alternatively solved by pushing away or tilting the test apparatus' were pooled.

3.2   Cognitive Task: Pipe

As for the board test, the proportion of solved trials (Table 3) did not differ between the environments (chi² = 0.46, p = 0.97). Similarly, non-enriched sows tended (p = 0.09) to solve the task faster than sows of the other environments (Table 4 and 5), while latencies for first touch were not influenced by housing. In contrast to the previous test, treatment had no influence on latency to first touch. However, the significant treatment x trial interaction showed a decrease in latencies solved (first trial test: 49.8 ± 17.8 vs first trial control: 34.8 ± 20.4 and second trial test: 34.8 ± 21.0 vs second trial control: 38.5 ± 19.7) and first touch-solved for the test pigs in the second trial (first trial test: 48.5 ± 19.5 vs first trial control: 33.1 ± 21.1 and second trial test: 30.4 ± 23.5 vs second trial control: 37.2 ±20.6). Post-hoc comparisons of the significant treatment x trial interaction showed no further differentiation.

Table 4: Latencies in s (Mean±SEM) to first touch the apparatus for test (n=18) and control (n=18) animals, by housing environment. Data from all animals (N=36) were included. When the task was not solved within the maximum limit of 60 sec, latency was set to 60 sec.

Cognitive test	Non-Enriched		Enriched		Super-Enriched		Significance of effects		
								p-values	
Board	Test	Control	Test	Control	Test	Control	Treatment	Environment	Trial
Latency First Touch (s)[1]	3.5 ± 2.0	9.0 ± 9.3	12.3 ± 12.9	25.1 ± 26.6	5.1 ± 2.8	13.8 ± 13.4	0.21		
							0.07 (NE vs E)	0.06 (NE vs SE)	0.81 (E vs SE) 0.11
Average	6.1 ± 1.5		18.6 ± 4.5		9.5 ± 2.8				
Pipe									
Latency First Touch (s)[1]	4.2 ± 4.1	2.5 ± 0.8	3.7 ± 1.2	4.3 ± 5.2	4.3 ± 2.9	2.5 ± 0.5	0.08		
							0.94 (NE vs E)	0.87 (NE vs SE)	0.90 (E vs SE) 0.62
Average	3.4 ± 0.8		4.0 ± 0.8		3.4 ± 0.5				

Table 5: Number of solved tasks (N) and latencies in s (LSMeans±SEM) from entering the arena until solving the task (Latency Solved) and from first touching the test apparatus until solving the task (Latency First Touch – Solved), by test and control animals and housing. Only data from solved trials were included.

Board

Traits	Non-Enriched		Enriched		Super-Enriched		Significance of effects p-values				
	Test	Control	Test	Control	Test	Control	Treatment	Environment	Trial	Treatment* Environment	Treatment*Trial
N Solved (S/AS)¹	10(6/4)	10(6/4)	9(1/8)	8(4/4)	9(0/9)	9(6/3)					
Latency Solved (s, N=)	19.1 ±4.7ᵃ	27.0 ± 5.3ᵇ	25.3 ± 4.1ᵃ	29.1 ± 6.0ᵇ	20.1 ± 3.0ᵃ	27.0 ± 4.5ᵇ	0.03	0.07	0.06	0.99	0.39
Latency First Touch –Solved (s)	16.3 ± 4.6	18.8 ± 3.7	15.6 ± 3.5	15.0 ± 2.4	15.0 ± 3.1	15.7 ± 3.6	0.26	0.37	0.07	0.99	0.53

Pipe

Traits	Test	Control	Test	Control	Test	Control	Treatment	Environment	Trial	Treatment* Environment	Treatment*Trial
N Solved (S/AS)¹	6(6/0)	7(6/1)	6(6/0)	8(6/2)	5(4/1)	9(8/1)					
Latency Solved (s)	15.0 ± 4.6	25.1 ± 5.7	26.8 ± 4.3	23.8 ± 6.7	24.8 ± 4.7	26.1 ± 3.3	0.21	0.09	0.20	0.48	0.04
Latency First Touch – Solved (s)	7.8 ± 3.0	22.8 ± 5.7	21.5 ± 3.4	21.6 ± 5.3	21.2 ± 5.3	23.8 ± 3.3	0.22	0.12	0.16	0.43	0.03

¹ Solution categories: S = solved by pulling rope; SA= alternatively solved by pushing away or tilting the test apparatus
a,b: means with different small letters within one row differ significantly by treatment (p<0.05)

4. Discussion

The aim of this study was to determine whether housing environments with varying levels of enrichment had an influence on the pigs' ability to solve a cognitive task. Further, it was investigated if animals which had previously been exposed to novel object tests (test animals) solved the cognitive task more easily than their naïve counterparts (control animals).

Both tests applied proved to be suitable for a short term test. Although animals had no previous learning training, about 67 % of all animals solved the tasks correctly within very short latencies. Solving percentage in the pipe test nearly doubled with the second trial, indicating quick learning capacities of the pigs and suggesting that the success rate would have increased with more frequent exposure. Interesting differences were observed between both tests. The solving strategy in the board test included more alternative solutions such as pushing away or tilting the test apparatus (in 44% of tests). This strategy persisted at a high level even in the second trial. However, in the pipe test pulling the rope was the most suitable strategy to retrieve the food item and only about 7% other solving reactions were observed. Apparently, the pipe test presented a different and possibly higher cognitive challenge than the board test.

One main finding of the study was that the test animals solved the board task faster than the naïve controls. For the control animals, finding the test apparatus upon entering the test arena was the first contact with a novel object in the test arena. This could have led the control sows to explore the test apparatus more cautiously and therefore also solve the task slower than the test animals. For the initiation of exploratory behaviour, which in our study was a necessary prerequisite for solving the task, novelty is an important factor (Berlyne, 1960). When presented with a choice between a familiar vs a novel object, piglets choose the novel object and also explored the novel object significantly longer than the familiar object (Wood-Gush and Vestergaard, 1991). Also in the study of Gifford et al. (2006), pigs preferred a novel object over a familiar one if they were given the opportunity to explore the familiar object for two days prior to the exposure of the novel object. Interestingly, our control sows decreased their latency to solve the pipe test in the second trial as shown by the significant treatment x trial interaction. This finding suggests an increased explorative motivation in the control animals compared to the test animals which were more habituated to novel objects.

The three housing environments with varying levels of enrichment had no significant effect on the test results. However, animals originating from the NE environment tended to solve both tasks faster than the pigs coming from enriched pens. This is in contrast to the findings of Mendl et al., (1997), where enriched pigs focussed more on finding the food quickly while non-enriched pigs were more interested in investigating the surroundings of the maze. Also in the study of Grimberg et al. (2015), enriched pigs took less time to find hidden food and showed less re-visits to un-baited holes in a holeboard task than barren housed controls. A possible explanation of the increased speed in solving both tasks by the NE animals could be a post-inhibitory rebound effect (Kennedy, 1985) which has been described for several animals species (i.e. rabbits Dixon et al., 2010, hens: Nicol, 1987, or horses: McGreevy and Nicol, 1998). Early ethological studies revealed that the motivation to perform a behavioural pattern increases during times at which this pattern cannot be carried out. As soon as the animal is given the opportunity to perform said behavioural pattern, the frequency or speed with which the behaviour will be shown will be larger or faster (Lorenz, 1950). This rebound effect could be the reason for the increased motivation of the NE animals to solve the task faster than the E and the SE sows.

The pipe test represented a more complex task than the board test and it was expected that housing environments with varying levels of enrichment would influence the pigs' ability to solve a cognitive task. It is tentatively suggested that the NE environment still offered a broad range of stimulating structures so that reduced behavioural flexibility and consequently lower learning and problem-solving abilities as suggested by the findings of Rosenzweig et al. (1964) were not induced. For the first four weeks of life the piglets in our study lived in commonly used barren pens. After weaning, the piglets were relocated into our experimental NE environment, which was barren apart from a thin layer of sawdust to avoid skin lesions. This management measure might have provided sufficient enrichment to counteract the possible impact of impoverished housing on the learning and problem solving behaviour of the NE pigs in our study.

Insights into the effects of housing conditions on the animals' mental state could potentially point to management procedures which include the animals' cognitive abilities in order to create living conditions which enhance welfare and minimise stress (Mendl et al., 2010). Enabling animals to make use of their cognitive abilities can also be considered as a form of enrichment in otherwise barren conditions. In the study of Zebunke et al. (2013), the effect of cognitive enrichment on the welfare of pigs was investigated. Experimental pigs were conditioned to discriminate an acoustic signal and thereupon receive a food reward, while control pigs were fed conventionally. After the conditioning phase, heart rate measurements revealed that upon the feeding announcement, the heart rate of both experimental and control pigs was rising. During feeding, however, the heart rate of the experimental animals decreased and their heart rate variability increased. The conventionally fed control animals showed a continued rise of their heart rate and a decrease in their heart rate variability during feeding (Zebunke et al., 2013). As high heart rate variability can be an indicator for positive emotions, the authors concluded that the cognitive enrichment of the experimental pigs led to relaxed feeding and a positive emotional state (Zebunke et al., 2013). Also the work of Puppe et al. (2007) confirms influences of cognitive enrichment on young pigs (7-20 weeks of age). Experimental animals were conditioned to an acoustic cue and had to perform an operant task (push a button) to receive food while control animals were fed conventionally. Observations in the housing environment as well as through open field and novel object tests revealed that pigs from the experimental group showed more locomotor behaviour, less fear behaviour and less belly nosing compared to the control group (Puppe et al., 2007).

We conclude from our findings that the level of enrichment in the home environment had an effect on the motivation of pigs to explore and interact with the test apparatus used in this study. The pigs coming from the non-enriched home environment tended to solve the cognitive tasks faster which is probably not due to differences in cognitive capacities but rather a rebound effect in motivation to show species specific exploration behaviour. This finding underlines the importance for pigs, having evolved as highly opportunistic animals, to explore and interact with their surroundings.

Cognitive enrichment is an important tool to create situations which allow pigs to have a certain degree of control over their environment combined with the opportunity to gain a reward. For the improvement of pig welfare, this is a win-win situation as it satisfies species-specific needs. Furthermore, the cognitive tasks used in this study are easily carried out, they do not require complicated technical equipment and could be also used under field conditions. If the apparatuses are baited with varying rewards, the degree of novelty remains high for the animals. As enrichment tools, they can therefore potentially be used to increase the welfare of pigs on farm.

References

Berlyne, D.E. 1960. Conflict Arousal and Curiosity. McGraw-Hill: New York, USA

Bolhuis, J.E., Oostindjer, M., Hoeks, C.W.F., De Haas, E.N., Bartels, A.C., Ooms, M., Kemp, B. 2013. Working and reference memory of pigs (sus scrofa domesticus) in a holeboard spatial discrimination task: the influence of environmental enrichment. Anim. Cogn. 16, 845-850.

Broom, D.M., Sena, H., Moynihan, K.L. 2009. Pigs learn what a mirror image represents and use it to obtain information. Anim. Behav. 78, 1037-1041.

Croney, C.C., Adams, K.M., Washington, C.G., Stricklin, W.R. 2003. A note on visual, olfactory and spatial cue use in foraging behaviour of pigs: indirectly assessing cognitive abilities. Appl. Anim. Behav. Sci. 83, 303-308.

D'Eath, R.B., Turner, S.P. 2009. The natural behaviour of the pig. In: Marchant-Forde JN (ed) The Welfare of Pigs pp 13-45. Springer, Netherlands.

Focardi, S., Capizzi, D., Monetti, D. 2000. Competition for acorns among wild boar (Sus scrofa) and small mammals in a Mediterranean woodland. J. Zool. 250, 329-334.

Frädrich, H. 1974. A Comparison of Behaviour in the Suidae. In: Proceedings of the Symposium on the Behaviour of Ungulates and its relation to management pp 133-143. International Union for the Conservation of Nature and Natural Resources.

Gaston, W., Armstrong, J.B., Arjo, W., Stribling, H.L. 2008. Home Range and Habitat Use of Feral Hogs (Sus scrofa) on Lowndes County WMA, Alabama. In: Proceedings of the National Conference on Feral Hogs. Paper 6.

Gieling, E.T., Nordquist, R.E., Van der Staay, F.J. 2011. Assessing learning and memory in pigs. Anim. Cogn. 14, 151-173.

Park, S.Y., Nordquist, R.E., Van der Staay, F.J. 2012. Cognitive performance of low- and normal-birth-weight piglets in a spatial hole-board discrimination task. Paed. Res. 71, 71-76.

Grimberg-Henrici, C.G.E., Vermaak, P., Bolhuis, J.E., Nordquist, R.E., Van der Staay, F.J. 2015. Effects of environmental enrichment on cognitive performance of pigs in a spatial holeboard discrimination task. Anim. Cogn. 19, 271-283.

Grundlach, H. 1968. Brutfürsorge, Brutpflege, Verhaltensontogenese und Tagesperiodik beim Europäischen Wildschwein (Sus scrofa L.). Ethol. 25, 955-995.

Haskell, M., Wemelsfelder, F., Mendl, M.T., Calvert, S., Lawrence, A.B. 1996. The effect of substrate-enriched and substrate-impoverished housing environments on the diversity of behaviour in pigs. Behav. 133, 741-761.

Held, S.D.E., Byrne, R.W., Jones, S., Murphy, E., Friel, M., Mendl, M.T. 2010. Domestic pigs, Sus scrofa, adjust their foraging behaviour to whom they are foraging with. Anim. Behav. 79, 857-862.

Kittawornat, A., Zimmermann, J.J. 2010. Toward a better understanding of pig behavior and pig welfare. Anim. Health Res. Rev. 1-8.

Lind, N.M., Moustgaard, A. 2005. Response to novelty correlates with learning rate in a go/no-go task in Göttingen Minipigs. Neur. Plast. 12, 341-345.

McLeman, M.A., Mendl, M.T., Jones, R.B., Wathes, C.M. 2005. Discrimination of conspecifics by juvenile domestic pigs, Sus scrofa. Anim. Behav. 70, 451-461.

Mendl, M.T., Laughlin, K., Hitchcock, D. 1997a. Pigs in space: spatial memory and its susceptibility to inference. Anim. Behav. 54, 1491-1508.

Mendl, M.T., Ehrhardt, H.W., Haskell, M., Wemelsfelder, F., Lawrence, A.B. 1997b. Experience in substrate-enriched and substrate-impoverished environments affects behaviour of pigs in a T-maze task. Behav. 134, 643-659.

Mendl, M.T., Held, S., Byrne, R.W. 2010. Pig cognition. Curr. Biol. 20, R196-R798.

Nielsen, T.R., Kornum, B.R., Moustgaard, A., Gade, A., Lind, N.M., Knudsen, G.M. 2009. A novel spatial delayed non-match to sample (DNMS) task in the Göttingen minipig. Behav. Brain Res. 196, 93-98.

Pearce, G.P., Paterson, A.M. 1993. The effect of space restriction and provision of toys during rearing on the behaviour, productivity and physiology of male pigs. Appl. Anim. Behav. Sci. 36, 11-28.

Puppe, B., Ernst, K., Schön, P.C., Manteuffel, G. 2007. Cognitive enrichment affects behavioural reactivity in domestic pigs. Appl. Anim. Behav. Sci. 105, 75-86.

Rosenzweig, M.R., Bennett, E.L., Krech, D. 1964. Cerebral effects of environmental complexity and training among adult rats. J. Comp. Physiol. Psychol. 57, 438-439.

Stolba, A., Baker, N., Wood-Gush, D.G.M. 1983. The characterization of stereotyped behaviour in stalled sows by informational redundancy. Behav. 87, 157- 182.

Stolba, A., Wood-Gush, D.G.M. 1989. The behavior of pigs in a semi-natural environment. Anim. Prod. 48, 419-425.

Van der Staay, F.J., Schoonderwooerd, A.J., Stadhouders, B., Nordquist, R.E. 2016. Overnight Social Isolation in Pigs Decreases Salivary Cortisol but Does Not Impair Spatial Learning and Memory or Performance in a Decision-Making Task. Front. Vet. Sci. 2, 81.

Van de Weerd, H.A., Docking, C.M., Day, J.E.L., Avery, P.J., Edwards, S.A. 2003. A systematic approach towards developing environmental enrichment for pigs. Appl. Anim. Behav. Sci. 84, 101-118.

Wemelsfelder, F., Haskell, M., Mendl, M.T., Calvert, S., Lawrence, A.B. 2000. Diversity of behaviour during novel object tests is reduced in pigs housed in substrate-impoverished conditions. Anim. Behav. 60, 385-394

Wood-Gush, D.G.M., Vestergaard, K. 1991. The seeking of novelty and its relation to play. Anim. Behav. 42, 599-606.

CHAPTER V

GENERAL DISCUSSION

GENERAL DISCUSSION

The current study aimed to investigate ethological and physiological indicators for positive emotional states in pigs. Further, the influence of different novel objects and varying levels of enrichment in housing environments on the expression of positive emotions was tested. Lastly, the role of positive emotions (through receiving a reward) with regards to learning behaviour was investigated. Emotions can be described by three components (Paul et al., 2005): 1. the ethological, 2. the physiological and 3. the conscious (felt) response to sensory input. The experiments conducted in this study (Chapters II, III and IV) focused on the measurement of the first two components. In the following chapter, the conscious component of emotion and its role with regards to animal welfare will be discussed. Further, the effects tested in this study (novelty, enrichment in housing environments and learning behaviour) and the relevance of positive emotions for animal welfare will be evaluated for farm animals. In a broader approach, also zoo- and laboratory animals will be considered.

Animal Consciousness

Irrespective of species, consciousness is commonly defined on two levels. Firstly, phenomenal or feelings consciousness ascribes awareness of feelings, sensations, thoughts and emotion to an individual (Block, 1998; Macphail, 1998). Secondly, self-consciousness describes the subjective awareness of ones-self as a thinking and feeling individual (Macphail, 1998; Damasio, 2000). The study of consciousness, however, is highly elusive and even human consciousness is largely not understood (Dawkins, 2012).

The physiological and ethological component of emotion entails major similarities between humans and non-humans (Dawkins, 2015) and these components were also studied in this dissertation. The third component of emotion however, consciousness, holds a widely discussed problem. In the study of human emotions, conscious feeling of the test person is commonly transferred verbally (Paul et al., 2005). This approach is not fruitful when experiments are conducted with animals, as they cannot verbally provide information on how they feel (Paul et al., 2005). Whether non-human animals have conscious experiences is not yet known (Mendl & Paul, 2004). The belief that non-human animals have conscious experiences is an essential determinant of animal welfare, but it is based on an assumption and therefore represents a weakness in animal welfare science (Mendl & Paul, 2004; Sandøe et al., 2004). As a result, scientists differ widely in their attitudes towards animal consciousness. As early as 1871, it was obvious for Charles Darwin that animals are sentient beings. As the continuity of evolution produced similar ethological and physiological reactions in humans and in animals i.e. in dangerous situations, also the accompanying conscious experience must be similar (Darwin, 1871 and 1872; Cabanac, 1971). Recent work using electrical and chemical stimulation of the brain (ESB and CSB, respectively) indicates that neural circuits generating primary affective feelings in mammals are indeed comparable (Panksepp, 2011). When specific regions of mammal brains are stimulated (chemically or electrically), distinctly similar emotional responses are observed for the specific species (Panksepp, 2011). Emotional responses entail vocalizations specific for the affectively intense situation mammals are in and are based on the similarity of activated brain structures (Jürgens, 2009).

Through brain research, it is known that the induction of positive emotional states through electrical or chemical stimulation can cause play behaviour to occur (Panksepp, 1981; Panksepp et al., 1985). As explorative behaviour is a very central behaviour to pigs (Studnitz et al., 2007), it might not be surprising that there could be a connection to a positive affective state. In the present study, explorative behaviour as well as play behaviour were observed in the test animals during novel object tests (Chapter II, III). By measuring cardiovascular parameters during tests, the aim was to find accompanying physiological indicators of positive emotional states indicated by low HR and high HRV. This was in parts found when pigs' explored complex objects (leaves) compared to more rigid objects (duck; Chapter II). Interestingly, vascular parameters indicative of positive emotional states could be seen to a larger extent when the animals were engaged in exploratory behaviour compared to when they were showing play behaviour. Play behaviour was observed mostly with the objects i.e. Kong® and rope and least with the complex objects leaves or soil (Chapter II). Play behaviour is usually associated with a higher degree of locomotor activity and arousal which might have caused a shift in the cardiovascular parameters as found in the present study. In the study of Zebunke et al. (2013), agonistic interactions among pigs resulted in high arousal during feeding and also caused elevated HR and decreased HRV in conventionally fed control pigs, compared to experimental pigs. Feeding itself might also have induced a positive affective state, which was superimposed by the arousal situation the animals were in (Zebunke at al., 2013). The present association between positive emotional state and exploratory behaviour in pigs merits further research. In this respect situations in which pigs are given the opportunity to explore complex objects or substrates could be carried out with more detailed measurement of physiological indicators of positive emotions in addition to established indicators such as HR and HRV (e.g., chemical or electrical brain stimulation). It is important to take into account influencing factors such as i.e. diurnal rhythms or agonistic interactions.

In rats, electrical stimulation of the brain (ESB) in specific regions mediating reward reliably resulted in 50 kHz ultrasonic calls and self-stimulation behaviour of the animals. Further, when a rewarding substance was administered, increases of 50 kHz calls were observed (Burgdorf et al., 2007). In the attempt to reproduce the 50 kHz calls in rats, the animals were stimulated in body regions which are also targeted during conspecific social play which resulted in enormous increases of the 50 kHz vocalization (Panksepp, 2007). Research into infantile human laughter by means of brain imaging shows that it is homologous in its neural base and function to the rats 50 kHz calls (Scott and Panksepp, 2003; Panksepp, 2007).

Going back to definition, these studies provide support for the phenomenal consciousness or the primary processes of the brain which are shared by all mammals. As primary processes center around the "tools for living provided by evolution", they incorporate the basic emotional systems (seeking, lust, care and play on the positive side; rage, fear and panic on the negative; Panksepp 2011). Secondary processes entail memory, learning and the unconscious functioning of the brain while the tertiary processes allow higher functions of the mind (Panksepp, 2011). What the above described studies as well as the present study measure however, are the ethological and physiological correlates of affective states, still assuming that i.e. laughing or playing consciously "feels" similar for humans and non-human mammals. Conscious experience and what a brain is doing is not easily connected, and varying states or levels of consciousness might depend on different neuronal processes (Stoerig, 2007; Morsella et al., 2010). Human consciousness (even though verbal transfer is possible) still remains a hard problem (Chalmers, 1996), more so because it becomes more and more apparent that large parts of human behaviour function unconsciously (i.e. driving or playing an instrument; Paul et al., 2005, Rolls, 2014), which might also be the case in animals (McPhail, 1998; Shettleworth, 2010). According to De Waal (2005), if animals show similar ethological reactions in specific emotional situations as humans, their accompanying conscious experience must be similar as well. However, if humans show behaviour through unconscious pathways, maybe this is also the case for animals (McPhail, 1998). Dawkins (2001, 2012) consequently suggested that if animals behave like humans, this could say less about their consciousness than commonly thought. With regards to the present study, this could mean that explorative behaviour could also happen in a neutral state of mind, because it might also be an automated behaviour, not necessarily associated with any emotional valence.

Studies on consciousness using brain imaging (PET, fMRI) also lead into controversial directions. According to McPhail (1998) language marks the fundament of consciousness and so only humans are conscious. Baars (2005) on the other hand concludes that even though language is absent in other species, homologies in brain structures, perception and memory are evident in all mammals and even reptiles suggesting that consciousness is also biologically fundamental in non-human species. Language indeed seems to play a subordinate role in consciousness. In cases of severe damage of linguistic regions of human brains, patients still consciously experience sensations like smells or touch but also emotions like joy or sorrow (Mendl and Paul, 2004), leading to the proposition that subjective feelings must be alike in humans and non-human animals, irrespective of language (Damasio, 2000). Mendl and Paul (2004) suggest investigating cognitive functions of animals which directly relate to consciousness in humans. In a study of Hampton (2001), two rhesus monkeys were trained in a delayed matching to sample task and the delay between the sample image and the four match images was steadily increased. At this stage of the experiment, the monkeys had to take the test. In a later stage of the experiment, the monkeys were given a choice whether they wanted to take the test or not. Taking the test and matching the sample image correctly resulted in a high quality reward (a peanut). Taking the test and incorrectly matching the sample resulted in no reward. If the monkeys decided not to take the test at all, a low quality reward was given (pellet feed). It was found that the monkeys were able to assess the quality of their memory. If the delay was too long and the likelihood of matching the sample image correctly was low, the monkeys decided not to take the test. It was concluded that the monkeys therefore know what they know (Hampton, 2001).

Held et al. (2000) paired two pigs (one lighter in weight than the other) in groups and set up a food search task. The lighter pigs were sent into the maize individually at first and therefore had the chance to locate the baited bucket out of the un-baited ones. When the lighter, now informed, pigs were sent into the maize again, they relocated the baited buckets faster. When sent into the maize together, the heavy, un-informed pigs followed the informed pigs and investigated the same buckets which were targeted by the informed pigs immediately before. Another result was that over half of the heavy, un-informed pigs exploited their informed partners by displacing them from the baited food buckets, consuming the food themselves. They therefore used the knowledge of their peers (Held et al., 2000). Crossing intra-species communication, Miklosi et al., (2000) were able to show that dogs use "showing" gestures in their communication with humans. Test dogs were shown food or a toy being hidden in an inaccessible location. When the owner was present, the dogs showed gaze alternation behaviour and kept looking from the owner to the hidden food and back. Such enhanced communication abilities in dogs are useful tools in order to obtain resources and are likely to be the result of selective breeding over the course of domestication (Miklosi et al., 2000). In a study by Nawroth et al. (2015), also dwarf-goats showed their ability to correctly differentiate between human body postures in a food anticipation paradigm. The goats were further able to use human point and touch cues in order to successfully infer the location of a reward (Nawroth et al., 2015).

Relationships between cognition and emotional state are multidimensional (Paul et al., 2005). From human studies, it is known that cognitive tests can induce a positive emotional state (Mathews and McLeod, 2002) and underlying positive emotional states can influence learning outcomes in cognitive tasks (Hinde, 1985; Forgas, 2000). In the cognitive tasks used in the present study (Chapter IV), it was hypothesized that the underlying emotional state of the animals coming from enriched housing environments would be more positive and their capacities to solve cognitive tasks would be superior to that of their barren housed conspecifics. However, on the contrary, it was found that the animals of the barren environment tended to touch the test apparatus faster and also tended to solve the tasks faster. The animals which previous to the cognitive tasks underwent novel object tests were used to encountering an object in the test arena, whereas the controls were not. The controls took longer to solve the tasks than the experimental animals, presumably because of their lack of experience and not because of their underlying emotional state (Chapter IV). Insights into the association between emotional state and learning behaviour in pigs were given by Carey and Fry (1993, 1995). In their study, pigs were adminstered an anxiolytic drug and trained to show an operant response in this state. Then they were trained to give a different operant response in their normal, un-drugged state. When then confronted with anxiety-inducing situations (i.e. transportation or exposure to an unfamiliar pig), the pigs showed the operant response previously associated to the drugged state and provided information to their emotional state.

The belief that animals are capable of consciously experiencing feelings and sensations, and therefore intrinsically deserve ethical considerations, still represents a cornerstone of animal welfare (Dawkins, 2015). Mental health gives animal welfare its moral weight and physiological well-being only, does not guarantee good welfare (Dawkins, 2015). However, intervention strategies for animal welfare problems like i.e. self-mutilation or injuries from unfitting housing environments can be developed irrespective of whether knowledge on animal consciousness exists or not (Wuerbel, 2009). Poor animal welfare (i.e. occurrence of stereotypic behaviour) can for example be improved through the provision of environmental enrichment (Mason et al., 2007). Physiological well-being (the absence of pain and disease) as well as mental health (contentment and pleasure) is therefore both a pre-requisite for good welfare (Wuerbel, 2009). Good welfare, the absence of chronic stressors and positive emotion is often marked by play behaviour (Held and Spinka, 2011; Boissy et al.,

2007) and, in case of the present study, also the ability to perform species specific behaviour.

According to Broom (2014), cognition describes the ability to have a perception of something in its absence and subjective feelings are not ending with this perception. So, most likely, cognition involves subjective feelings or consciousness (Broom, 2014). Irrespective of our knowledge on the animals' conscious experience of positive emotion, physiological functions are measurable and remain an important foundation in the improvement of animal welfare as long as empiric studies on animal consciousness continues to be out of scope of modern scientific methods. The notion that animals do not experience conscious feelings or do not have a definition of 'self' has no greater evidence based truth-value than the assumption that they do (Panksepp, 2011). Hence, as long as there is no evidence against animal consciousness, animals should be given the benefit of the doubt (Bradshaw, 1998). Results of the present study in this context highlight, that the effect of underlying emotional state induced by housing enrichment on cognitive performance is not necessarily straightforward.

Practical implications for farm and zoo animals

In contrast to livestock animals that are farmed for only a relatively short period of time until they have reached slaughter weight or their production of milk or eggs decreases, zoo animals for instance often face many years of confinement in surroundings which do not offer the space or complexity like the animals respective natural habitats. As a result, zoo animals as well as farmed animals show a range of abnormal behaviours (Brion, 1964; Meyer-Holzapfel, 1968; Fraser, 1975). In order to establish what abnormal behaviour is, extensive knowledge on the biology and ethology of animals is needed (Broom and Fraser, 2015). Normal behaviour is observed in complex environments where animals have the opportunity to express their full behavioural range (Broom and Fraser, 2015). Abnormal behaviours in livestock animals includes stereotypic behaviour like pacing and weaving (horses), crib biting (horses), bar-biting (pigs), excessive grooming (calves), tongue-rolling (cattle), sham chewing (sows), wool-pulling (sheep), inter-sucking (calves), tail biting (pigs), snout-rubbing (pigs) and drinker-pressing (pigs) (Broom and Fraser, 2015). In zoo animals, abnormal behaviours vary according to taxonomic group. While carnivorous animals rather display abnormal behaviours of locomotor nature, ungulates often develop oral stereotypies (Mason, 2003; Mason, 2006). Behaviours like self-biting, over-grooming, eating inedible objects, pacing and rhythmic rocking are not uncommon in zoo animals (Meyer-Holzapfel, 1968; Templin, 1993).

The terms 'stereotypical behaviour' and 'abnormal behaviour' have different meanings. Commonly defined, stereotypies are considered 'repetitive, unvarying, functionless behaviour patterns' (Mason et al., 1991). However, there are for instance stereotypies like over-grooming that involve variable motor patterns (Mills and Luescher, 2006; Garner, 2006). Or there are behaviours which seem functionless, but are unproblematic and perfectly normal for the species (e.g. pillow-kneading in cats; Mason et al., 2007). Consequently, it has been suggested to reserve the term 'stereotypic behaviour' to repetitive behaviours "induced by frustration, repeated attempts to cope and/or CNS dysfunction" (Mason, 2006). In zoo animals however, the reason for stereotypic behaviour is not always known as there is a big variety of species and not every case is empirically investigated (Mason, 2006). Therefore, the term 'abnormal repetitive behaviour' (ARB) is the best alternative (Garner, 2006); whereas 'stereotypic behaviour' refers to a shown behaviour that is definitely caused by deficits in housing environment leading to frustration (Mason et al., 2007).

Mason et al. (2006) suggested that ARBs can indicate environments which cause negative emotional states in animals. Such environments are aversive because they offer low stimulation and a high level of physical confinement (Clubb and Mason, 2002). Also social isolation (Novak et al., 2006) and unavoidable stress or fear (Meyer-Holzapfel, 1968; Novak et al., 2005) was found to be the causes leading to highest prevalence of stereotypic behaviours and ARBs. However, stereotypies and ARBs might not always reflect the current suboptimal housing conditions but may be a "scar" from previous inadequate housing environments (Swaisgood and Shepherdson, 2005). The expression of stereotypies and ARBs is a way of coping with aversive environments, leading to the suggestion that animals who do show these behaviours have better welfare than those that do not. Still, 68% of environments in which animals show stereotypic behaviour are associated with poor welfare conditions (Mason and Latham, 2004).

The zoo community itself was on the forefront of developing environmental enrichment methods in order to address the problem of stereotypic and abnormal behaviour in captive wild animals and develop strategies for improvement (Swaisgood and Shepherdson, 2005).

Supplementing impoverished housing systems with additional resources is often termed 'environmental enrichment' (Newberry, 1995). It is argued, that a "truly" enriched environment however, entails more than just the satisfaction of basic needs through the regular enrichment of an already varied environment with resources or cognitive means (Duncan and Olsson, 2001; Boissy et al., 2007). In the present study therefore, only the animals housed in the super-enriched environment lived in truly enriched circumstances, as the already enriched environment (sawdust and straw) was regularly supplemented with additional resources (hay, grass cut, twigs, leaves and vegetables) (Chapter III, IV). Findings in literature suggest that novelty of resources elicit an increase in play behaviour in pigs and, if given a choice, pigs will choose novel over familiar objects (Wood-Gush and Vestergaard, 1991). This was also found in the present study, as the novel objects presented to the experimental animals elicited more play and tail wagging behaviour compared to the behaviour of the controls, which hardly showed any play or tail wagging behaviour in the test arena. Novel objects might hold the potential to induce positive emotional states if presented in a way that is not sudden and predictable. This was also shown by Désiré et al. (2004) in lambs. When confronted with a novel object, the lambs reacted with a startle response and an increase in HR upon the sudden presentation of the object. When the objects were not presented unexpectedly, the lambs orientated towards the novel object and reacted with an increase in HRV (Désiré et al., 2004). For the present study, it would have been interesting to observe the behaviour of the pigs in their home environment. Especially the comparison of the play behaviour frequency between the NE, E and SE animals might have revealed whether the regularly added additional resources in the SE environment had an effect on play behaviour (Chapter III, IV).

Conclusions

The main aim of the present thesis is to investigate potential behavioural indicators of positive emotion in young pigs in combination with underlying physiological parameters. It was hypothesized that behavioural expressions of positive emotions (i.e. play and tail wagging) are accompanied by physiological changes (increased HRV, decreased HR) indicative of positive emotional state in response to novel objects.

It was found that there were some relations between exploratory behaviour and cardiovascular parameters. When the animals were expressing behavioural traits suggesting positive emotional states (such as play and tail wagging) the measured cardiovascular parameters were not coherent. This might have been caused by increased locomotor activity during play. Even though not for all tested novel objects a coherent explanation of accompanying cardiovascular parameters was found, it was interesting to observe that pigs spent significantly more time exploring more complex objects as opposed to more rigid novel objects. As cardiovascular parameters are easily influenced by confounding factors, more detailed research into the association of exploratory behaviour and positive emotional states in pigs is merited, especially taking the influence of locomotor activity and diurnal rhythm on heart rate into account.

Pigs that were confronted with novel objects expressed more play and tail wagging behaviour than control pigs, which did not have the opportunity to explore novel objects. The behavioural responses of the experimental pigs expressed a positive underlying emotional state, leading to the assumption that the novel object tests were positively valenced. Also, increased arousal and locomotor activity in the experimental animals might have caused high HR and low HRV, limiting conclusive interpretations of the association between behavioural and physiological traits and their relation with emotional states. Pigs housed in impoverished conditions had a higher positive emotion score (PES) in the behaviour tests than their enriched housed conspecifics. This finding could hint toward a rebound effect, describing the need for stimulation in animals housed in barren conditions which is discharged to a greater extent in situations where exploration and stimulation is given. In this context, it would be interesting to compare cardiovascular parameters in the test situation and home pen situation in order to elucidate emotional states which are caused by novelty and the base level due to the housing environment in pigs.

Rebound effects for exploration and stimulation were also observed in the cognitive test of this study. Animals housed in the barren environment tended to approach the test apparatus and solve the tasks faster than their enriched conspecifics. This finding highlights the intrinsic need for pigs to explore. The test animals (who had previous experience with novel objects) solved the board task faster than the naïve control animals, while parallel differences were not apparent in the pipe task. Interestingly, the naïve animals decreased their latency to solve the pipe task in the second trial. This finding suggests an increased explorative motivation in the control animals compared to the test animals which were more habituated to novel objects. Novelty is therefore also an important influencing factor in cognitive tasks. Overall, cognitive enrichment serves as an opportunity for pigs to control their environment and also gain a reward and therefore have the potential to induce positive emotional states.

References

Baars, B. J. 2005. Subjective experience is probably not limited to humans: The evidence from neurobiology and behavior. Conscious. Cogn. 14, 7-21.

Boissy, A., Manteuffel, G., Jensen, M. B., Moe, R. O., Spruijt, B., Keeling, L. J., and Bakken, M. 2007. Assessment of positive emotions in animals to improve their welfare. Physiol. Behav. 92, 375-397.

Burgdorf, J., Wood, P. L., Kroes, R. A., Moskal, J. R. and Panksepp, J. 2007. Neurobiology of 50-kHz ultrasonic vocalizations in rats: electrode mapping, lesion, and pharmacology studies. Behav. Brain Res. 182, 274-283.

Bradshaw, R. H. 1998. Consciousness in non-human animals: adopting the precautionary principle. J. Conscious. Stud. 5, 108-114.

Brion, A. 1964. Les tics chez les animaux. *Psychiatrie Animale* (eds. A. Brion and H. Ey) Desclee de Brouwer, Paris, 229-306.

Broom, D.M. 2014. Cognition in Relation to Emotion. In: Broom, D.M. (Ed.) Sentience and Animal Welfare. CAB International, Wallingford.

Broom, D.M. and Fraser, A.F. 2015. Domestic Animal Behaviour and Welfare, 5th Ed. CAB International, Oxfordshire, UK

Cabanac, M. 1971. Physiological role of pleasure. Science 173, 1103-1107.

Chalmers, D. J. 1996. The conscious mind: In search of a fundamental theory. Oxford University Press.

Clubb, R., Mason, G. 2002. A review of the welfare of zoo elephants in Europe. Horsham, UK: RSPCA.

Damasio, A. 2000. The feelings of what happens. Body, Emotion and the Making of Consciousness, Vintage, London, UK.

Darwin, C. 1871. The descent of man and selection in relation to sex. Reprinted by Princeton University Press (1981).

Darwin, C. 1872. The expression of emotion in man and animals. Reprinted by Chicago University Press (1965).

Dawkins, M. I. S. 2001. Who needs consciousness? Anim. Welf. 10, 19-29.

Dawkins, M. S. 2012. Why animals matter: animal consciousness, animal welfare, and human well-being. Oxford University Press, UK.

Dawkins, M. 2015. Animal Welfare and the Paradox of Animal Consciousness. Adv. Stud. Beh. 47, 5-38.

Désiré, L., Veissier, I., Després, Gérard, Boissy, A. 2004. On the Way to Assess Emotions in Animals: Do Lambs (Ovis aries) Evaluate an Event Through Its Suddenness, Novelty, or Unpredictability? J. Comp. Psych. 118, 363 – 374.

De Waal, F. 2005. Animals and suspicious minds. New Scientist, 2502, 48.

Forgas JP. 2000. Introduction: the role of affect in social cognition. In: Forgas, J.P (Ed.) Feeling and thinking: the role of affect in social cognition. Cambridge: CUP, 1-28.

Fraser, D. 1975. The effect of straw on the behaviour of sows in tether stalls. Anim. Sci. 21, 59-68.

Garner, J. P. 2006. Perseveration and stereotypy: systems-level insights from clinical psychology. In: Mason, G. and Rushen, J. (Eds.) Stereotypic Animal Behaviour. Fundamentals and Applications to Welfare, 2ª Ed. CAB International, Wallingford.

Hampton, R. R. 2001. Rhesus monkeys know when they remember. Proc. Natl. Acad. Sci. 98, 5359-5362.

Held, S., Mendl, M., Devereux, C., and Byrne, R. W. 2000. Social tactics of pigs in a competitive foraging task: the 'informed forager'paradigm. Anim. Behav. 59, 569-576.

Held, S. D., and Špinka, M. 2011. Animal play and animal welfare. Anim. Behav. 81, 891-899.

Hinde R.A. 1985. Was the 'expression of the emotions' a misleading phrase? Anim. Behav. 33, 985-92.

Jürgens, U. 2009. The neural control of vocalization in mammals: a review. J. Voice, 23, 1-10.

Macphail, E. M. 1998. The evolution of consciousness. Oxford University Press, UK.

Mason, G. 1991. Stereotypies: a critical review. Anim. Behav. 41, 1015-1037.

Mason, G. 2003. Captivity effects on wide-ranging carnivores. Nature 425, 473-474.

Mason, G. J., Latham, N. 2004. Can't stop, won't stop: is stereotypy a reliable animal welfare indicator? Anim. Welf. 13, 57-69.

Mason, G. 2006. Stereotypic behaviour in captive animals: fundamentals and implications for welfare and beyond. In: Mason, G. and Rushen, J. (Eds.) Stereotypic Animal Behaviour. Fundamentals and Applications to Welfare, 2ª Ed. CAB International, Wallingford.

Mason, G., Clubb, R., Latham, N., and Vickery, S. 2007. Why and how should we use environmental enrichment to tackle stereotypic behaviour? Appl. Anim. Behav. Sci. 102, 163-188.

Mathews A, McLeod C. 2002. Induced processing biases have causal effects on anxiety. Cogn. Emot. 16, 331-54.

Mendl, M. and Paul, E. S. 2004. Consciousness, emotion and animal welfare: insights from cognitive science. Anim. Welfare 13, 17-25.

Meyer-Holzapfel, M. 1968. Abnormal behavior in zoo animals. In: Fox, M.W. (ed.) Abnormal behavior in animals. W.B. Saunders, Philadelphia, Pennsylvania, 476-503.

Miklósi, A., Polgárdi, R., Topál, J., Csányi, V. 2000. Intentional behaviour in dog-human communication: an experimental analysis of "showing" behaviour in the dog. Anim. Cogn. 3, 159-166.

Mills, D., Luescher, A. 2006. Veterinary and pharmacological approaches to abnormal repetitive behaviour. In: Mason, G., Rushen, J. (Eds.) Stereotypic animal behaviour: fundamentals and applications to welfare, 2nd ed. 325-356. Cab International, Wallingford.

Morsella, E., Krieger, S. C., Bargh, J. A. 2010. Minimal neuroanatomy for a conscious brain: Homing in on the networks constituting consciousness. Neural Networks, 23, 14-15.

Nawroth, C., von Borell, E., Langbein, J. 2015. 'Goats that stare at men': dwarf goats alter their behaviour in response to human head orientation, but do not spontaneously use head direction as a cue in a food-related context. Anim. Cogn. 18, 65-73.

Novak, M. A., Meyer, J. S., Lutz, C., Tiefenbacher, S. 2006. Social deprivation and social separation: developmental insights from primatology. In: Mason, G. and Rushen, J. (Eds.) Stereotypic Animal Behaviour. Fundamentals and Applications to Welfare, 2ª Ed. CAB International, Wallingford.

Novak, M. A., Meyer, J. S., Lutz, C., Tiefenbacher, S. 2005. Stress and the performance of primate stereotypies. In: Mason, G. and Rushen, J. (Eds.) Stereotypic Animal Behaviour. Fundamentals and Applications to Welfare, 2ª Ed. CAB International, Wallingford.

Paul, E., Harding, E., Mendl, P. 2005. Measuring emotional processes in animals: the utility of a cognitive approach. Neurosci. Biobehav. Rev. 29, 469- 491.

Panksepp, J. 1981. Brain opioids – A neurochemical substrate for narcotic and social dependence. In: Cooper, S. (Ed.) Theory in Psychopharmacology. Academic Press, London.

Panksepp, J., Jalowiec, J., DeEskinazi, F.G., Bishop, P. 1985. Opiates and play dominance in juvenile rats. Behav. Neurosci. 99, 441-453.

Panksepp, J. 2007. Neuroevolutionary sources of laughter and social joy: Modeling primal human laughter in laboratory rats. Behav. Brain Res. 182, 231-244.

Panksepp, J. 2011. The basic emotional circuits of mammalian brains: do animals have affective lives? Neurosci. Biobehav. Rev. 35, 1791-1804.

Paul, E. S., Harding, E. J. and Mendl, M. 2005. Measuring emotional processes in animals: the utility of a cognitive approach. Neurosci. Biobehav. Rev. 29, 469-491

Rolls, E. T. 2014. Emotion and decision making explained. Oxford, Oxford University Press.

Sandøe, P., Forkman, B. and Christiansen, S. B. 2004. Scientific uncertainty—how should it be handled in relation to scientific advice regarding animal welfare issues? Anim. Welfare 13, 121-126.

Scott, E., and Panksepp, J. 2003. Rough-and-tumble play in human children. Aggressive Behav. 29, 539-551.

Shettleworth, S. J. 2010. Clever animals and killjoy explanations in comparative psychology. Trends Cogn. Sci. 14, 477-481.

Studnitz, M., Jensen, M.B., Pedersen, L.J. 2007. Why do pigs root and in what will they root? A review on the exploratory behaviour of pigs in relation to environmental enrichment. Appl. Anim. Beh. Sci. 107, 183-197.

Stoerig, P. 2007. Hunting the ghost: Toward a neuroscience of consciousness. The Cambridge handbook of consciousness, 707-730.

Swaisgood, R. R., Shepherdson, D. J. 2005. Scientific approaches to enrichment and stereotypies in zoo animals: what's been done and where should we go next? Zoo Biol. 24, 499-518.

Templin, R. 1993. Stereotypic movements in zoo animals. In: Proceedings of the International Congress on Applied Ethology, Humboldt University, Berlin, Germany, 54-59.

Würbel, H. 2009. Ethology applied to animal ethics. Appl. Anim. Behav. Sci. 118, 118-127.

Zebunke, M., Puppe, B., Langbein, J. 2013. Effects of cognitive enrichment on behavioural and physiological reactions of pigs. Physiol. Behav. 118, 70-79.

Danksagung

Ich danke Prof. Dr. Martina Gerken, die mir diese Position ermöglicht hat, für den Vorschlag dieser interessanten Thematik, ihren Rat und ihre Unterstützung während der Entwicklung meiner Dissertation.

Prof. Dr. Ute Knierim dafür, dass sie sich als Zweitprüfer zur Verfügung gestellt hat und Prof. Dr. Achim Spiller, der sich bereit erklärt hat, als Drittprüfer mitzuwirken.

Dem Ministerium für Wissenschaft und Kultur des Landes Niedersachsen für die Förderung des Projekts.

Dem Promotionsprogramm Animal Welfare, was tiefere Einblicke und Perspektiven ermöglicht hat, und dadurch sehr viel zu meiner professionellen Entwicklung beitragen konnte.

Prof. Dr. Uta König von Borstel, für die Leihgabe der Herzfrequenz-Messgeräte.

Dr. Ahmad Reza Sharifi, für die Unterstützung bei der Entwicklung statistischer Analysen.

Birgit Sohnrey, für ihre Geduld und Hilfsbereitschaft bei meinen Laborarbeiten.

Prof. Dr. Czerny und den Mitarbeitern der Abteilung Mikrobiologie und Tierhygiene für die Möglichkeit der Probenbearbeitung und Auswertung. Besonderer Dank gilt hier Caroline Bierschenk, Henrike Ahsendorf und Thomas Kinder für die geduldige Anleitung, die Beantwortung meiner vielen Fragen und die Auswertung meiner Proben im Labor.

Jürgen Dörl, Dieter Daniel und Knut Salzmann für die technische Hilfe und die Versorgung meiner Versuchstiere, die ich stets in besten Händen wusste und deren Arbeit ich sehr hochachte. Hier auch ein besonderer Dank für die mentale Unterstützung und Kraft an den Tagen an denen die Tiere abgeholt wurden.

Meiner Arbeitsgruppe. Vielen Dank an Alex Riek für ein offenes Ohr, Rat und Tat in allen Promotionsfragen. Lea Brinkmann für eine harmonische Bürozeit, das Lesen meiner Texte und die Hilfe in praktischen Angelegenheiten wenn ich diese brauchte. Vivian Gabor, für eine Verbundenheit im Geiste, fachlichen Rat und tolle Gespräche. Verena Hauschildt, für eine schöne gemeinsame Zeit am DNTW und für den hilfreichen Input zu meinen Papern. Rhuksana Amin Rhuna, worries are only half, when shared. Thank you for your kind support and our conversations. I think we took it all with a good sense of humour. Friederike Warns und Thomas Köhn, für die Begleitung der Endphase meiner Dissertation mit viel Witz und vielfältigen Mensa-Gesprächen.

Diemut Labusch, die mich in der Endphase meiner Dissertation enorm gestärkt hat.

Meinen Eltern, Christine und Gary McKenna, die immer Platz ließen für meine Neugierde an allem Lebendigen, die mir Zugang zu Büchern verschafften, die mir meine akademische Ausbildung ermöglichten und mich doch in erster Linie nicht deshalb lieben.

Meinem Mann, Sergej Prinz, für seine Unterstützung. Fürs Mut machen, wenn meiner mich verlassen hat, fürs an mich glauben, wenn ich das nicht mehr konnte.

Meinem Sohn, Tom McKenna Prinz, dafür, dass du mich nach draußen entführst und wir nach Regenwürmern suchen. Damit bringst du alles in die richtige Perspektive.

Lebenslauf

Persönliche Daten

Name: McKenna

Vorname: Lisa Christine

Geburtsdatum: 06.04.1987

Geburtsort: Pirmasens

Ausbildung

10/2013 – 02/2018

Promotion zum Doktor der Agrarwissenschaften, Department für Nutztierwissenschaften, Georg-August-Universität Göttingen, 37075 Göttingen, Deutschland

09/2011 – 09/2013

Masterstudium Animal Science, Studienschwerpunkt: Tiergesundheit und Tierverhalten, Wageningen University, Wageningen, Niederlande

09/2007 – 07/2011

Bachelorstudium Equine. Leisure and Sports, Van Hall Larenstein University of Applied Science, Wageningen, Niederlande

2006

Abitur Wirtschaftsgymnasium Pirmasens, Deutschland

Praktika

2010

Forschungsassistenz Equine Research and Information Centre, Macclesfield, United Kingdom

Göttingen, September 2018

www.ingramcontent.com/pod-product-compliance
Lightning Source LLC
Chambersburg PA
CBHW060320220326
41598CB00027B/4381